FEATURES

SUMMER 2022 • NUMBER 32

Plough

ARTS & LETTERS

WEB EXCLUSIVES

Read these articles at *plough.com/web32*.

Plough
PLOUGH.COM

EDITOR: Peter Mommsen
SENIOR EDITORS: Maureen Swinger, Sam Hine, Susannah Black
EDITOR-AT-LARGE: Caitrin Keiper
MANAGING EDITORS: Maria Hine, Dori Moody
BOOKS AND CULTURE EDITOR: Joy Marie Clarkson
POETRY EDITOR: A. M. Juster
DESIGNERS: Rosalind Stevenson, Miriam Burleson
CREATIVE DIRECTOR: Clare Stober
COPY EDITORS: Wilma Mommsen, Priscilla Jensen
FACT CHECKER: Suzanne Quinta
MARKETING DIRECTOR: Trevor Wiser
UK EDITION: Ian Barth
CONTRIBUTING EDITORS: Leah Libresco Sargeant, Brandon McGinley, Jake Meador
FOUNDING EDITOR: Eberhard Arnold (1883–1935)

Plough Quarterly No. 32: Hope in Apocalypse
Published by Plough Publishing House, ISBN 978-1-63608-055-0
Copyright © 2022 by Plough Publishing House. All rights reserved.

EDITORIAL OFFICE
151 Bowne Drive
Walden, NY 12586
T: 845.572.3455
info@plough.com

SUBSCRIBER SERVICES
PO Box 8542
Big Sandy, TX 75755
T: 800.521.8011
subscriptions@plough.com

United Kingdom
Brightling Road
Robertsbridge
TN32 5DR
T: +44(0)1580.883.344

Australia
4188 Gwydir Highway
Elsmore, NSW
2360 Australia
T: +61(0)2.6723.2213

Plough Quarterly (ISSN 2372-2584) is published quarterly by Plough Publishing House, PO Box 398, Walden, NY 12586.
Individual subscription $32 / £24 / €28 per year.
Subscribers outside the United Kingdom and European Union pay in US dollars.
Periodicals postage paid at Walden, NY 12586 and at additional mailing offices.
POSTMASTER: Send address changes to Plough Quarterly, PO Box 8542, Big Sandy, TX 75755.

Photograph on page 100: installation view, *Vasily Kandinsky: Around the Circle*, Oct. 8, 2021–Sept. 5, 2022, Solomon R. Guggenheim Museum, New York. Photograph by David Heald © Solomon R. Guggenheim Foundation, New York.

ABOUT THE COVER:
Jeremy Collins's detailed artwork *The Storm* evokes the tempest of current events that threatens to overwhelm us from above and below. But the traveler in the boat holds on in hope that it will carry him safely to harbor. See more of Collins's artwork at *jercollins.com*.

FORUM ≈
LETTERS FROM READERS

This Forum features responses to *Plough's* Spring 2022 issue, "Why We Make Music." Send contributions to *letters@plough.com*, with your name and town or city. Contributions may be edited for length and clarity and may be published in any medium.

DRENCHED IN GRACE

It was the worth the cost of subscription simply to find out, in Stephen Michael Newby's "Go Tell It On the Mountain," that Duke Ellington once held a concert of sacred music, capped off by a tremendous tap performance by Bunny Briggs as the orchestra played "David Danced before the Lord with All His Might." That has to be seen to be believed. To quote from another of the edition's many fine pieces, some things "exceed our ability to praise them" ("Dolly Parton Is Magnificent," by Mary Townsend). To read this magazine, and to be pointed to the things its words give glimpses of, is to be drenched in grace. Our world is starved for grace, and for gratitude; in response to all of the former that flows from your contributors' work to your readers, please receive the deepest of mine of the latter.

Kevin Sunday, Mechanicsburg, Pennsylvania

AMATEUR BACH

On Maureen Swinger's "Doing Bach Badly": Swinger begins her piece with the comment: "When our amateur choir sings Bach's *Saint Matthew Passion*, the music's power overwhelms our mistakes." Having listened to – and even joined – Bruderhof choirs singing hymns after supper, or the *Messiah* in Advent, countless times over the past twenty-seven years, I was momentarily startled by the word "amateur." I have always found the multi-part harmonies of Bruderhof singers a peak experience of musical transport, directly to my most receptive soul. But of course, this music *is* amateur, proceeding from love and not from profession.

Peggy R. Ellsberg, Ossining, New York

A SOUTHERN LADY

On Mary Townsend's "Dolly Parton Is Magnificent": Kalon is one foreign word that seems to summarize Dolly Parton's career. I've often thought of Arabic *nabil*, which used to be inadequately translated as "largesse" and "honor." (Yes, this is what some Southerners mean by talking of a Southern lady's "honor.")

The idea is that the purpose of having money is to share it, use it to reward good and punish evil – the "greater" your wealth and fame, the more you give and share. The concept seems to have been widespread in the ancient world; without Christian (or even Islamic) values as a guide the

idea could deteriorate into ridiculous extravagance, as in the Roman feasts. Even with Christian values it could be misguided – rich Victorians scraping all the leftovers off all the plates into a bucket and having it sent to "the poor" rather than fed to the pig. Still, it's an idea we could do with a little more of.

I grew up with so many memories of Dolly Parton and her music. Dad was an instant fan after hearing "Coat of Many Colors" and pushed me to sing "Love Is Like a Butterfly." Mother thought Dolly looked too trashy but admitted some of her songs are great. Everybody here goes to Dollywood for day trips, with dates or with children. But I think what I'll remember best will be the way she set up a flash fund for survivors of the 2016 forest fire on the Tennessee–North Carolina border. While many disaster survivors were sitting back and whining for government bailouts, Dolly Parton just called a few friends and started funding. That is what Real Southern Ladies do.

Priscilla King, Gate City, Virginia

INCLUSIVE MUSIC

On Adora Wong's "How to Make Music Accessible": As to what makes music "inclusive," I am reminded of the "think system" in *The Music Man*: one does not have to hear the right notes in precisely the right order to appreciate another's attempt to make joyful noise; the beauty is as much in the ear of the listener. I am also reminded of my daughter with special needs, playing on the keyboard, singing along to a CD

Artwork by Marta Zamarska. Used by permission.

she is playing and later transcribing it to a sheet full of joined half notes. She is entirely convinced that she has done something worthwhile, although only a very few could appreciate it. Her mother and I do.

Martin Bohley, Palmyra, New Jersey

BIBLICAL LULLABIES

On Norann Voll's "How to Lullaby": My wife and I have been developing a repertoire of lullabies for our newborn son that we began singing when he was in the womb. We don't only sing religious songs, but the songs that resonate most with me are those adapted from scripture that I learned from my own parents when I was a child:

Give ear to my words, O Lord
Consider my meditation
Hearken unto the voice of my cry . . .

As Brittany Petruzzi points out, people throughout history have found spiritual strength from singing the psalms. Even Jonah, in the fish's belly, reorients his life to God by singing psalms.

Although the joy that our little one brings us makes the struggle in raising him worth it – and more – there are moments, such as on a long day after a sleepless night of trying to settle him down, when I feel like I am in a sea storm or even drifting in the belly of a fish. Thanks to singing such spiritual lullabies, like Jonah, I too can pray amid the waves and breakers sweeping over me.

Joshua Seligman, Broughton-in-Furness, England

CAN ANABAPTISM BE CATHOLIC?

On the November 2021 commemoration in Vienna "Let Brotherly Love Remain" with Cardinal Christoph Schönborn, Heinrich Arnold, and others: I was delighted to read the report on the commemoration of the early Anabaptists in Vienna. It is a powerful fruit, in my view, of the Second Vatican Council in which so many of the theological formulations of the Radical Reformers (the universal call to holiness, the priesthood of all believers, the importance of returning to the scriptures to develop doctrine, to name a few) were definitively embraced by the Catholic Church. It is my constant prayer to see that healing continue.

It is common to see these ecumenical efforts as dialogues of "separated brethren" with an assumption that the separation is permanent, that the best we can expect is to get along decently going forward. Jacob sent his gifts to Esau, but they never again became one nation. But I feel an urgency when I read the Lord's parting words, "that they may be one." Titled after these very words, *Ut unum sint*, an encyclical of Saint Pope John Paul II on ecumenism, suggests that "from this basic but partial unity it is now necessary to advance toward the visible unity which is required and sufficient and which is manifested in a real and concrete way, so that the churches may truly become a sign of that full communion in the one, holy, catholic, and apostolic church which will be expressed in the common celebration of the Eucharist." Rather than once-fighting but forgiven brothers, is it possible to imagine a future where we are at last one body, the Bride of the Lord Jesus?

The possibilities for this kind of corporate reunion seem more abundant now than ever. Since the sixteenth century, more than a dozen Eastern churches have united with Rome, but nonetheless maintain their independent jurisdiction and particular traditions. As a Roman Catholic, I have been deeply enriched by their liturgies, prayers, and witness to the gospel, especially through my spiritual mentor, a Byzantine Catholic nun. In 2009, the Traditional Anglican Communion's request to unite with the

Marta Zamarska, *Accordian Player 1*, oil on canvas, 2014

ABOUT US ≈

Plough is published by the Bruderhof, an international community of families and singles seeking to follow Jesus together. Members of the Bruderhof are committed to a way of radical discipleship in the spirit of the Sermon on the Mount. Inspired by the first church in Jerusalem (Acts 2 and 4), they renounce private property and share everything in common in a life of nonviolence, justice, and service to neighbors near and far. There are twenty-nine Bruderhof settlements in both rural and urban locations in the United States, England, Germany, Australia, Paraguay, South Korea, and Austria, with around 3000 people in all. To learn more or arrange a visit, see the community's website at *bruderhof.com*.

Plough features original stories, ideas, and culture to inspire faith and action. Starting from the conviction that the teachings and example of Jesus can transform and renew our world, we aim to apply them to all aspects of life, seeking common ground with all people of goodwill regardless of creed. The goal of *Plough* is to build a living network of readers, contributors, and practitioners so that, as we read in Hebrews, we may "spur one another on toward love and good deeds."

Plough includes contributions that we believe are worthy of our readers' consideration, whether or not we fully agree with them. Views expressed by contributors are their own and do not necessarily reflect the editorial position of *Plough* or of the Bruderhof communities.

FORUM ≈
CONTINUED

Catholic Church was granted, resulting in the growing "Anglican Ordinariate." In these cases, what has occurred is not a matter of conversion to Catholicism, but a genuine *reunion*.

If I may make so bold, I pray that there one will be one day be an "Anabaptist Ordinariate." Living in Lancaster, my faith has been constantly nourished by your tradition. It was the testimony of a Bruderhof couple at an intentional community retreat that inspired my wife and me to begin our marriage in such a community. Soon after, our Catholic Worker House was co-founded by a "Mennonite Catholic," received into the latter church but continuing to draw from his roots in the former. Now, I talk often with my Amish coworkers about our concordances, from the sacred chant in our liturgies to the Luddite critique in our traditions. While the sacraments and social teaching keep me rooted in the Catholic Church, I frequently suggest to my fellow parishioners that we need to be living more like Anabaptists if we want to take the gospel seriously (as I slide a copy of *Plough* into their hands). Again, I pray that my fellow Catholics among the laity will avail themselves of your community's spiritual wisdom, and that our leaders might be eager to offer self-jurisdiction and the integration of your heritage, if ever the opportunity of the "visible unity" of "full communion" arrives.

In this age of upheaval, who knows what new things Christ will call us to: "where sin abounds, grace abounds all the more." No matter what happens, or what we live to see, I find peace in the assurance that we will continue to pray for one another and the flourishing of our Christian communities.

Sean Domencic, Holy Family Catholic Worker, Lancaster, Pennsylvania

SINGING IN CHURCH

On Ben Crosby's "Is Congregational Singing Dead?": Thank you for this article! I am one who misses traditional music in the church. I am a faith and cultural Mennonite whose family has been in Canada since 1876. I miss the beautiful four-part harmony, either a capella or accompanied; I miss the hymnbooks and the discipline in singing they provide. I struggle with the "new music." It doesn't encourage harmony and sometimes the performance goes on and on (and on and on . . .) I want the comfort of tradition, but I also need community.

Lois Thiessen, Winnipeg, Manitoba

Congregational singing has diminished because the modern worship songs, played by a rock band with a solo female singer, are entirely unsuitable for congregational singing, not least because they are pitched too high for men's voices and don't have four-part harmony settings which men would be able to sing. There's something badly wrong with a choice which has the consequence of making Christian men unable to sing worship in church.

Angus Johnson, Kent, England ⟿

FAMILY& FRIENDS
AROUND THE WORLD

Photograph courtesy of Citlalli Ríos

Diaconía Paraguay

One family's call to mission evolves into a microloan and employment organization.

Coretta Thomson

In 2009, Citlalli Ríos felt God calling her to study theology. When she brought up this revelation to her husband Roberto Hernández, she was amazed to hear that he had been hearing the same call. They and their two young children traveled over four thousand miles from their home in Ixtapaluca, a suburb of Mexico City, to the Mennonite seminary in Asunción, Paraguay. They expected to return home four years later to their busy church life, Roberto's microfinance job with World Vision, and the house they had almost finished paying off. But within two years, their plans changed. They

Coretta Thomson is a Plough *editor and a member of the Bruderhof.*

crossed paths with Judah Mooney, who wanted to start a Christian microfinance organization in Paraguay. The three of them founded Diaconía Paraguay, which eventually obtained the helpful backing of the Global Aid Network.

In 2021 alone, Diaconía helped over five thousand families to start food and handicraft businesses. Citlalli leads the educational division of the mission, coordinating more than forty free workshops annually. Last year, two thousand people learned how to make bread, wrap thermoses with leather, and decorate hats with bead-jewelry bows, among other skills. Diaconía also provides pastoral services through volunteers lead "trust groups." Some have found Christ through these relationships, while others have simply found hope and support. When someone is going through a difficult time

or falling behind on a loan repayment, everyone in the group helps out.

In March 2020, when Covid-prevention measures locked down the world, the effects were most severely felt in countries where a large proportion of the population lives hand-to-mouth – countries like Paraguay, and people like Diaconía's typical clientele. Citlalli encouraged the women she worked with to rise to the challenge with faith. Some diversified their businesses. Others signed up for Diaconía's video seminars, gaining skills through the long days of quarantine. In 2021, Diaconía members were invited to teach classes in a local women's prison, which eventually extended to the halfway house, where the guards and director joined in.

The fact that the stories of Diaconía participants can be summarized in a sentence belies the dramatic positive impact

A Diaconía baking workshop was held in a local school; the students enjoyed the results.

that a microloan, skill, and support network can have on the women and families involved. Runilda, a twenty-two-year-old mother of three, picked trash for a living until Diaconía helped her start a clothing store. Blanca, an incarcerated woman, has learned to crochet, decorate tiles, and adorn footwear with colorful beadwork. Not only has she earned money while in prison, but she plans to continue this work when she is free. Rosanna has used a series of microloans over the course of ten years to start and grow her business. After successfully selling jewelry, she moved on to clothing and small appliances. When the pandemic hit, she branched out into groceries.

Paraguay has been Citlalli and Roberto's home for thirteen years; their children are now young adults. They arrived on tickets financed through American friends, without any social or long-term financial support, but they have gained dozens of coworkers and friends. They are glad to be part of this mission, but, says Citlalli, "if the Lord calls us elsewhere, we're ready to go."

Winners of the second annual Rhina Espaillat Poetry Award

Congratulations to winner Bruce Bennett for his poem "Stopping by with Flowers" and to finalists Sherry Shenoda for "Sugarcane Memories" and Eric T. Racher for "Sonnet addressed to George Oppen, Arlington National Cemetery," all published in the pages of this issue. The winners were announced at a livestreamed event with Rhina P. Espaillat and *Plough* poetry editor A. M. Juster. The award is for a poem of not more than fifty lines that reflects Espaillat's lyricism, empathy, and ability to find grace in everyday events of life. The 2022 competition attracted over 750 poems.

Bruce Bennett was born in Philadelphia, Pennsylvania, in 1940. He received his AB, AM, and PhD from Harvard, and from 1967 to 1970 taught at Oberlin College, where he co-founded and served as an editor of *Field*. In 1970 he moved back to Cambridge, where he co-founded *Ploughshares*. In 1971, he married Bonnie Apgar, a Renaissance art historian; they have two children. In 1973 he began teaching at Wells College in Aurora, New York. At Wells, Bennett taught British and American literature and creative writing, was chair of the English department, and served as director of Wells College Press, which he helped to found. Bennett retired in June 2014.

Bennett is the author of ten full-length collections of poetry and more than thirty poetry chapbooks. His most recent book is *Just Another Day in Just Our Town* (Orchises, 2017). He was awarded a Pushcart Prize in 2012, and his poems have appeared widely in literary journals, textbooks, and anthologies. See his poem "Stopping By with Flowers" on page 31.

Sherry Shenoda is an Egyptian-American poet and pediatrician, born in Cairo and now living near Los Angeles with

her husband and two sons. She won the 2021 Sillerman First Book Prize for African Poets for her poetry collection *Mummy Eaters* (University of Nebraska Press, 2022) and was shortlisted for the 2019 Brunel International African Poetry Prize. She is also the author of a novel, *The Lightkeeper* (Ancient Faith, 2021).

Shenoda serves as a pediatrician in a nonprofit health center and has studied the effects of armed conflict on child health. She says her Coptic Orthodox Christian faith informs her practice of medicine, and her practice of medicine informs her faith, and both inform her writing. See her poem "Sugarcane Memories" on page 47.

Eric T. Racher was born in Akron, Ohio, and currently lives in Riga, Latvia. He is the author of a chapbook of poetry, *Five Functions Defined on Experience: For Jay Wright* (2021), and his work has appeared or is forthcoming in *Poetica*, *Dreich*, and *Maximus*. See his poem "Sonnet Addressed to George Oppen, Arlington National Cemetery" on page 51.

Plough's 2023 poetry competition is now open. The winner receives $2000, and two finalists receive $250. All three will be published in Plough. *Submit your new poems at* plough.com/poetryaward.

Searching for Safety

In a refugee center outside Vienna,
Ukrainian families wonder what comes next.

ROSALIND STEVENSON

Plough's graphic designer, Rosalind Stevenson, and her husband, Seth, spent the month of April in Neulengbach, Austria, working at a newly designated center for refugees from Ukraine. She reports:

Week One: This refugee center is a large complex of five buildings that have been mostly empty for the past few years. The first two buildings are now occupied; the others would need some fixing up before they could be lived in. There are approximately two hundred refugees here now, mostly women and children but also some men. A few of the men got out before it became mandatory to stay. Others were allowed to leave because they had disabilities or were fathers of large families. But most of the families we've had contact with had to leave their fathers and husbands behind.

When the first refugees arrived, there was an outpouring of support from the locality, in the form of donations and volunteers, but little oversight or organization. By the time we got here, a few weeks later, it was somewhat more organized, at least in terms of dealing with the donations, and the quantity has slowed down a little. Seth is setting up shelving to store it all, and putting donated furniture away,

Artwork by
a Ukrainian
refugee child
at the center in
Neulengbach

often working with the older boys and men from the center.

Upstairs, two rooms have been set aside for the kids to use, with donated furniture, toys, and games. But no one was organizing any activities for them, and with their fathers absent and their mothers overwhelmed, they would mostly roam around unsupervised or sack out with their phones. So I now spend the mornings doing projects and crafts with whoever is interested. Sometimes we'll see them at the end of the day on a Zoom call with their dads, showing them the art projects they made. The kids also have online classes during the day at certain times – some of their teachers stayed behind and are now conducting classes remotely for their scattered students.

Week Two: We got our outdoor play area up and running this week, with a swing set, slide, sandbox, mini soccer, and a volleyball net, as well as a few kiddie picnic tables and some benches for the parents. It is being heavily used already, and when the weather is nice, we generally get a game going in the afternoon.

As we get to know more of the refugees individually, we hear more of their stories:

Irina[*] is a mother of two who has become a spokesperson and leader for the refugees at Neulengbach, since she's the only one who can speak German well and communicate their needs. She tells us how suddenly life changed – for the past eight years she had been focusing on raising her children and securing their education: then overnight the country was at war, and they had to flee their home. Now she has almost no time for relaxing or being with her kids, as she is run off her feet sorting out the hundreds of questions that arise every day. She does everything from administering Covid tests to enrolling kids in school, helping people with documentation, and being a go-between with

* Names have been changed throughout this article.

the mayor and local officials. She gets as little as two hours of sleep a night.

Irina has asked me if I could start helping one mom who has six young children and is never managing to get out of the little apartment they're in. The kids are crammed in, four to a room in very small spaces, and sometimes spend all day in there. The mom doesn't know many people at the center and has no real support network, so I have been taking one of the children, a two-year-old girl, out for a walk and some outdoor play each day.

Katerina is a mother who has started taking English lessons with me. She comes from Lviv, and her husband and twenty-five-year-old daughter are still there, defending their country. She talks to them often, but they aren't allowed to give out any details, just confirmation that they're alive. She finds it very hard to think about anything except what is happening back home. Halfway through a lesson her phone buzzes, and she shows me that it is an air raid alarm in Ukraine. She signs with her hands that she is praying, and then wants to continue learning. Eventually an all-clear comes. But you can imagine how hard it is to

Children playing outside the center

focus on daily life when this is happening to family members. We tell her that we came all the way from New York because our church is concerned about the situation and wants to help, and she expresses her thanks.

Anya, a nineteen-year-old, comes from a town just outside Kyiv. She describes how, despite the rumors, no one believed that a war could happen in their peaceful home. It was only when they were woken by the sound of bombs early one morning that they were forced to accept the reality, and flee.

Week Three: Most of the school-aged Ukrainian children have started to attend the local school in Neulengbach. This is exciting for the kids and they are very happy, but there are challenges too. The obvious problem is that most do not speak any German, and their parents cannot help with homework or even communicate with the teachers. And of course, the kids need a lot of school supplies, lunchboxes, good shoes, pens and pencils. The school does provide some basics, but for much of it, they must rely on donations. I have great respect for the teachers who are finding ways to make a place of welcome.

Despite the language barrier, Seth has become friends with most of the men here, and one morning one of the fathers shows him a video someone just sent from a location near his home. The clip is the aftermath of a rocket attack on a train station in the town of Kramatorsk. When we check BBC later in the morning, we find that it is headline news, and that nearly fifty people have died there, most of whom were trying to flee the Eastern provinces. This and other similar incidents are reminders of the reality of the war, and how close it is to the people we're with every day.

For these families, there is no getting away from the war, even if they are physically safe at present. Husbands and fathers and other relatives are in constant danger; the future is completely unknown. Their most basic material needs might be addressed at the moment, but there is a lack of social workers, or someone to help them figure out the next step. Translation technology enables us to connect and hear their stories, albeit one sentence at a time. Each story makes us thankful for every child and family who has made it this far. But how many more are still in the center of a war zone? ⇝

War and the Church in Ukraine

*A pastor describes ministering
in wartime Bucha and Kyiv.*

IVAN RUSYN

Ivan Rusyn *(left)* and other church leaders serve open-air communion to Ukrainian soldiers.

Ivan Rusyn is the president of the Ukrainian Evangelical Theological Seminary (UETS) in Kyiv. He spoke with Plough's *Susannah Black by videoconference on April 9, 2022.*

Susannah Black: Dr. Rusyn, what's been the timeline of your war so far?

Ivan Rusyn: I returned from the States four days before the war started. February 24 is my wife's birthday; we planned to have dinner with her family. But that was the day everything started.

That first day, I was able to get to the seminary to implement our crisis plan. I went home that evening – we live in Bucha, just four or five miles from the seminary. By the next morning, the bridge between Bucha and Kyiv had been blown up, and there was fighting. We spent five days in the apartment building's basement gym, where I used to go for fitness. Then my wife evacuated with her two brothers and their five children; she stayed in West Ukraine with my mom for almost forty days. I returned to Kyiv, where I've been sleeping in a sleeping bag in one of the offices of the Bible Society of Ukraine.

In Kyiv, we have electricity, cell service, and internet, but in many places, including Bucha, there has been no telephone connection, no internet, no electricity for more than a month. That's one of the reasons I went back to Bucha at the beginning of March: as far as I knew, my neighbors were still in that basement, and there was supposed to be an evacuation corridor. But with no internet, the people had no idea.

So I got as far as I could by car, and then I took a bicycle and got back home with that. I told my neighbors and others about the evacuation corridors. Many of them were afraid to evacuate because they'd heard that Russians were killing civilians who were trying to escape.

On my way into Bucha, Ukrainian soldiers had stopped me. I told one, "I'm a priest – I take full responsibility for my actions." And he said, "I am very proud of you that you want to risk your life, but we will not allow you to go." But I was riding my bicycle – I found a way, and got across the river.

You snuck past the Ukrainian army?

Yeah. And then when I was coming back to Kyiv, I saw Russian soldiers, so I had to hide, but I managed to find a way out. At one point I was trying to cross a road and on the right and left hand were Russian military vehicles – not tanks, smaller ones, six of them.

Then I went back to Bucha again after it was liberated. We were able to drive, so we went with our minibus and delivered medicine and food. We've been back every day since then.

You always wonder how you would react, being in a war zone. What's surprised you?

Sometimes I cannot comprehend that this is happening. Two days ago our team visited Hostomel. You can't imagine how apocalyptic this picture is: You stop by an exploded building, and wait by the Red Cross trucks for two minutes, three minutes, and then one person shows up. And then one by one, other people come, mostly elderly. They are very

dirty. One lady noticed my surprise at how she looked. She said, "We are sorry about how we look; we've been cooking over open fires."

You don't ask people to share their stories; they just start. That lady told us that her husband was killed and she buried him in her backyard. She started to cry, and she hugged me. I've received more hugs from strangers since this began than from my family during the last five years.

People talk a lot about praying; they say God protected them. Yesterday or the day before – in this situation we don't remember days – I met a lady who lives in my apartment building. I'm there in my collar, a priest, a believer, and she tells me she's an atheist. Then she says, "I have a

UETS staff distribute food and necessities to the people remaining in Bucha.

special request for you." When she was in that basement, sheltering, there was a lady who had an icon and lost it. This other lady, the atheist, says to me, "Can you write to her that I found that icon? I will be happy to return it to her."

When I think about where God is in this, I don't have answers. But I was reading Elie Wiesel's book *Night*. He's sharing the story of when three Jews were murdered in the presence of other Jews, and somebody said, "Where is God now?" And Wiesel said that he heard a voice in his heart that God is killed. I feel the same when I see people killed, civilians and soldiers – I see Christ being killed.

"Where is God?" That is a question. But a few days ago I started to think that the proper question is not where is God, but where is humanity. This is not a question for God, but a question for us.

I have huge anger toward Russians, but also pity, because one day they will find out what was going on. I don't know how they will absorb that information because it seems like they're living in a different world. But this bubble will blow up; they will be exposed to reality.

Today we were delivering food in a smaller village, and people started to share stories. One elderly lady said that they were hiding in a basement and Russian soldiers entered. Her granddaughter screamed in fear, and Russian soldiers pointed a gun at the child. The lady said that for the three weeks since, the child has not been speaking. Such fear and such trauma. One day Russians will be exposed to this truth; I just don't know how they will live with it.

There is a song in Ukrainian which goes: "My killer calls me a sister." Russians want to say that we are brothers. Why do you kill us, then?

You talked in another interview about the imprecatory psalms, praying for God to break the teeth of your enemies. Have you been able to pray for them at all?

I try to be authentic and honest. I used to be a pacifist. When I was called up; I chose alternative national service. Now I believe that only the nation that has known the horror of war has the right to speak about pacifism. My theology has been changed. For me, peace-making is not a passive thing anymore, an ability to absorb and embrace everything. No, it is very active – action in order to stop violence.

When I ask God to intervene – to break the bones of my enemies – I know that his reaction will be proper, timely, and just. So it is about justice and about my recognition that we are absolutely dependent on God. When you compare the size of Russia and Ukraine, you will see that we are fighting a giant. The only hope we have is God. So, yes, I do pray. I don't pray about peace, I pray about victory. Peace will be an outcome of victory. Unfortunately, with Russia, there will be no peace without victory.

We have been serving communion for our soldiers in the open air. We say, "Thank you for your service." They say, "No, thank you for *your* service." The church is present; we haven't fled to somewhere else. And I think that after this war, many Christians, as well as secular people, will ask, "Where were you when we were being killed?" And Christian leaders will be able to say, "I was with you. I was here. I was in Kyiv." And it will be very powerful. So I think the church will be in a position to speak, and the voice of the church will be heard, because on the darkest days of our history, we stood together with the Ukrainian nation.

In the beginning of the war, maybe the seventh or eighth day, there was an ecumenical prayer in the very precious and important Saint Sophia Cathedral in downtown Kyiv. It's close to Ukraine's Security Service headquarters, so there was information that Russia might attack that area and the church, which is over a thousand years old, might be destroyed.

The All-Ukrainian Council of Churches decided to have a prayer inside: Catholics, Protestants, Orthodox, and also Muslims and Jews. We demonstrated to our society that we are here. It's about solidarity. I believe that the church will be stronger, will be more authentic, and will be an integrated part of our society. Our seminary is interdenominational. Our students are mostly Evangelical, but sometimes Orthodox, sometimes Catholic. Recently I've been visiting the Catholic archbishop of Kyiv. We have lunch, we share resources, we share food. It's a unique moment of unity.

In the midst of this, what gives you hope?

Well, one day you are full of hope, another you feel extreme emptiness. Every day we get stronger and stronger in our hope, but the first few days were very difficult.

When I see unjust suffering – that gives me hope. Because knowing something about God, I know that he will not be passive. He will intervene – he has already intervened. Someone asked me, "Do you still believe in God?" I may have had some doubts before the war, but now I have none.

How can our readers pray for you?

Pray for Ukraine. Pray for Ukrainian churches. One day this war will be over. I am convinced that only the church has the capacity to be the platform for healing, restoration, and, at some point, reconciliation. So my prayer is that the church in Ukraine will serve our wounded nation. And I think not only our theological or counseling expertise will help us: we will have the same scars; we went through the same suffering and this is what will make us able to serve our nation.

And also . . . I would like to ask you to pray for Russia. Because Russia is going nowhere. And this war already brings a lot of suffering to the Russian people. So pray that God will intervene and stop Putin so Russia will not be destroyed, so it will not fall apart economically.

We are not political institutions and the church cannot be a political power. But the church cannot be far from what is going on.

We try to have connections with Russians, our brothers and sisters, but it is very hard.

Jesus Christ was not a politician. But the statement "Jesus is Lord" has political resonance. This is about our ultimate loyalty. Worship has political resonance. Whom do we worship? The gospel is so powerful that it transforms every area of our life. We are not just waiting for evacuation to heaven.

If we are Christians, we have to have an impact. Yes, we are not of this world, but we are in this world for the sake of this world. We have to be engaged if we want to be a true church. For me it was very important that I remain here with my people. If I evacuate before everybody else, what kind of pastor am I?

When we shut the seminary [after it was bombed], it was very important to me that I be the one to close our campus gate. So I did it. I said, "I will be last." We are not political institutions and the church cannot be a political power. But the church cannot be far from what is going on. ⌖

For a longer version of this interview as well as Ivan Rusyn's wartime diary see plough.com/rusyn. *To support UETS's relief efforts, visit* uets.net/en/projects/relief-efforts.

Hoping for Doomsday

*The times are troubled. That's why we need
the promise of apocalypse.*

PETER MOMMSEN

WHERE THE EARTHWORMS STOPPED, a world had ended. The telltale sign was an eight-inch-deep layer in the geological record at a site in present-day Syria called Tell Leilan. As reported in *Nature* earlier this year, archaeologists digging there "found a buried layer of wind-blown silt so barren there was hardly any evidence of earthworms at work. . . . Something [had happened] that choked the land with dust for decades, leaving a blanket of soil too inhospitable even for earthworms."

The culprit, the archeologists concluded, was climate change: a century of drought 4200 years ago that plunged much of the inhabited world into chaos. Before, the region surrounding Tell Leilan had been the wheat-growing breadbasket for Mesopotamia's cities; then it became a dustbowl. The drought toppled the Akkadian Empire, one of history's first super-regional states, which collapsed amid famine and civil war. Formerly thriving urban centers, including the Tell Leilan site, were abandoned. A mass movement of climate migrants swept southward, where conditions were less severe; in a move many today may find familiar, the southern cities erected a border wall. It would be three centuries before Mesopotamia returned to a measure of stability. Meanwhile, the same drought, known as the 4.2 kiloyear event, seems to have devastated societies elsewhere too. It has been linked to civilizational collapse in the Indus River Valley and in Egypt, where the Old Kingdom succumbed to anarchy as the Nile's water level dropped by five feet.

What happened to the Akkadians could well be what's about to happen to us. The effects of climate change already show eerie parallels to the aftermath of the ancient drought. Newly unpredictable weather in Central America has caused crop failures and upended traditional agriculture, accelerating migration to cities and the United States. Supersize wildfires regularly devastate homes from Portugal to Australia to California. Extreme heatwaves in heavily populated areas of India and the Sahel have grown more frequent, with temperatures that may soon become lethal to human beings.

All this is only the beginning of what awaits us and our grandchildren, climate scientists warn, even if carbon emissions fall substantially. Nor does today's environmental crisis stop with weather. Biologists fear that animal and plant species are dying out at a rate comparable to that of the five mass extinctions that have occurred in the 3.7 billion years since life first appeared.

In the face of what many see as a mortal threat to life on earth, some have taken to extreme acts of protest. On Earth Day this year, Wynn Bruce, a fifty-year-old from Boulder, Colorado, burned himself to death outside the US Supreme Court, which was considering the legal validity of certain federal restrictions on carbon emissions. A friend and fellow Zen adherent told reporters that his decision to self-immolate was a "deeply fearless act of compassion" to protest government inaction on climate change.

Still, I can't help but wonder if Bruce wasn't also driven by a sense of dread and desperation that is far more widely shared. A worldwide 2021 study of sixteen- to twenty-five-year-olds

Ivanka Demchuk, *Archangel Michael*, mixed technique on canvas and wood, 2018

reports that four out of five young people are fearful of the future of the climate, with 59 percent "very" or "extremely worried." While a decade ago "eco-anxiety" was still a novel diagnosis, today psychologists are in increasing demand by those seeking therapy for debilitating fears about the threat to the planet.

More significantly in the long run, climate fears are one reason that people are having fewer children, at a time when most wealthy countries report a below-replacement fertility rate. For example, in a 2018 survey asking childless adults in the United States why they hadn't had children, 33 percent cited "worries about climate change" and 27 percent "worries about population growth." Whatever the reasons or circumstances for childlessness at an individual level, the society-wide retreat from childbearing runs counter to the most basic biological imperative: to bring the next generation into existence. Failing to do so seems to bespeak a despair about the future, for which plenty of reasons can be found beyond the climate.

Indeed, from another point of view, this pessimism may not be pessimistic enough. Concerns about global warming will prove superfluous in the event of nuclear war, a peril that few people under forty had spent long considering before Russia's invasion of Ukraine. Like a revenant from the Cold War, it has now returned as a real possibility. Millions of lives, and perhaps civilization itself, could be snuffed out in a matter of hours simply because of the desperation of one autocrat in Moscow. In that case, future archaeologists sifting through the geological layers of present-day London or Manhattan might well find radioactive dust and, presumably, no earthworms.

One way or the other, one day *Homo sapiens* will go extinct, with or without our help through carbon emissions or nuclear war, and the game will be over. At least that is what current scientific models foretell. Perhaps it

will be at the next round of global glaciation, predicted in a hundred millennia or so, but if not, the end will come when increased solar radiation kills off plant and animal life (perhaps in 600 million years), or at the latest when the sun balloons into a red giant and gobbles up the inner planets (7.5 billion years). The end of the solar system likely imposes a hard cutoff, even in the case that Peter Thiel's dreams of life extension work out or Elon Musk's plans for Martian colonies prove doable. Not even techno-futurism, it seems, can save us.

THE TRUTH, HOWEVER, is that few of us are *that* troubled by any of these end-of-the-world scenarios, even if we intellectually acknowledge them. That's why those who take them with a literalness we don't – self-immolating protesters; hammer-wielding peace activists denting nuclear submarines and then serving long prison sentences – may command a certain awe, but also repel. The attitude of most people, even the sympathetic, might be summed up by tweaking Augustine's prayer: *Lord, make me care about the end of the world, but not yet.*

Yet there is one kind of End Times that we can't avoid taking seriously sooner or later: the end of our personal world. Every child eventually learns that dying isn't something only other people do. For me the moment when this outrageous fact hit home found me in a dentist chair. The dentist was chiding me, a college freshman, about a broken tooth (lesson for the kids: don't open beer bottles with your canines). "You'll need these teeth another seventy years," she said, meaning to emphasize the long life that lay ahead of my jaw and me. Her words had the opposite effect. "Just seventy years?" I wanted to shout, though the dental dam made that impossible. Age nineteen is probably late for this kind of basic arithmetic to sink in, but it did then: just a few more decades for me, then the gravedigger scene from *Hamlet*. When I told my mother, a family physician, she responded as someone who has stood at dozens of deathbeds. "Time you realized," she said. "Flesh decays."

One response to the inevitability of the flesh decaying is a kind of nihilism. The blogger Freddie de Boer eloquently lays out this case:

> We're born in terror, we exist for no reason, we experience confusion and shame as children, we busily prepare ourselves for lives we don't want or can't have, we are forced to take on the burdens of adult responsibility, we compromise relentlessly on what life we'll pursue, we settle and settle and settle, we fear death and ponder our meaninglessness, we experience the horrors of aging, and when we die the only comfort we have is that we aren't conscious to learn that there was never any heaven or God to give it all meaning.

Looked at this way, meaninglessness is the fate even of good people who live enviable lives. Recently a woman I've known since early childhood died at age eighty-eight, surrounded by her children and dozens of grandchildren and great-grandchildren. She was a genuine matriarch, loved by hundreds of people beyond her family thanks to an extraordinary lifetime spent in outreaching care and service. A thousand people stood around her grave as we buried her. Yet much as she will live in our memories, as the old saying goes – and she truly will – all those who remember her will also die, and her memory with them. Even having many biological descendants doesn't count for much beyond a single lifetime – to one's grandkids' grandkids, after all, one will be only a very distantly related stranger.

"Teach me that there must be an end of me, and that my life has a finish, and that I must leave it," wrote the Psalmist, in words set to haunting music by the agnostic Johannes Brahms in his *Requiem*. "How utterly vain are all human beings, who live as if they were safe! They walk about like a shadow, and give themselves much trouble to no purpose; they gather up, and do not know who will get it." Brahms's music, which uses Luther's German translation of the Hebrew, culminates in the repeated, despairing question: "*Wes soll ich mich trösten?*" What comfort do I have to look to?

The paradoxical answer that ancient Judaism gave to such despair was a promise: the promise of doomsday. The coming "Day of the Lord" is a repeated theme of the Hebrew prophets, from Amos in the eighth century BC to the Book of Daniel in the second. (Similar prophecies appear in Zoroastrian scriptures around the same time.) On that day, the prophets declared, God would come to visit his people, taking vengeance on oppressors both foreign and homegrown and establishing lasting justice and peace. Wrath would be followed by renewal – for Israel, and perhaps for all of humanity as well, even the entire cosmos. The Book of Isaiah foretells how the natural world itself will be restored with the arrival of "new heavens and a new earth"; God will "swallow up death forever" and "wipe away the tears from all faces."

Isaiah is not speaking of "the end of the world," notes the scholar N. T. Wright. Rather, he is foretelling a future in which God's good creation, far from being destroyed, is transformed and renewed. This is what the "end of the age" meant to Jesus and his early followers, including the author of Revelation, who concludes his book with a majestic elaboration of Isaiah's prophecy. According to this vision, our own death is not the end after all, nor will scientific predictions of human extinction ultimately come to pass. "See, the home of God is among mortals. . . . He will wipe every tear from their eyes. Death will be no more; mourning and crying and pain will be no more, for the first things have passed away."

THIS HOPEFUL ANTICIPATION of the future isn't what most people associate with prophecies of doomsday. On the contrary, the Day of Judgment as commonly received in Western culture is primarily about terror – the *Dies Irae* whose woes are detailed in the Latin Mass for the Dead: "That day, day of wrath, calamity, and misery, day of great and exceeding bitterness, when Thou shalt come to judge the world by fire." In paintings such as Michelangelo's *Last Judgment* in the Sistine Chapel, it is the abject fear of those rising to be judged and the agony of the damned that seize the viewer's attention and remain longest in the memory, not the happiness of the blessed.

So it's understandable that when describing the threat of large-scale catastrophe, whether from climate change or nuclear war, people often end up reaching for a biblical word that evokes the end of the world most forcefully: *apocalypse*. Thus we have apocalyptic novels (from Mary Shelley's *The Last Man* to Walter M. Miller Jr.'s *A Canticle for Leibowitz* to Cormac McCarthy's *The Road*); apocalyptic movies (*On the Beach* to *Melancholia* to *Don't Look Up*); and apocalyptic sects (Jim Jones's People's Temple to the Branch Davidians to Aum Shinrikyo). For all the variety of these examples, the common thread of this way of thinking about apocalypse is: the end is nigh, and it will be bad.

But "apocalypse," in the New Testament, does not mean "the end of the world" any more than Isaiah's prophecies do. The English term derives from the first word in the last book of the Bible, the Revelation of John, where its meaning is "unveiling." What is being unveiled in this book? The world as it really is, and as it will be. Human events, in the original apocalyptic view, are more than just the ebb and flow of purposeless occurrences. What drives them forward is neither the individual will nor material determinism, but rather the battle of spiritual forces of good and evil, of which

political structures and social movements are merely manifestations. Not that the individual is powerless or unimportant – John's Apocalypse pointedly challenges its readers to choose which side of the cosmic struggle they will join, and rules out lukewarmness as an option ("Would that you were either hot or cold!"). Its claim is rather that there's a winnable war to wage. Revelation is thus one extended argument against nihilism; it insists that there *is* a God who gives it all meaning.

Admittedly, this book of unveiling is, in words of the scholar Christopher C. Rowland, "paradoxically the most veiled text of all in the Bible." Far from giving us a straightforward account, it offers instead a succession of cryptic sayings and fantastic images: angels, horsemen, plagues, many-headed beasts, a lake of fire, a cube-shaped city. Whatever sort of unveiling the book may represent, it isn't the disclosure of a timetable.

That hasn't stopped generations of pious futurists from trying to read it like one. Indeed, much of Christian history might be told as a story of calculations of doomsday dates, followed by inevitable disappointments. One of the reasons Augustine wrote his masterpiece *The City of God* in AD 426 was to counter a popular end-times theory focused on the year 500; when that year came and went, people set their hopes successively on 801, 1000, 1033, and so on, up to the present day. (According to the seventeenth-century Anglican divine James Ussher, whose biblically based chronology of the earth proved widely influential, the eschatological millennium would dawn in 1997 – just in time, had it happened, to prevent Vladimir Putin from becoming president of the Russian Federation.) In parallel to arithmetical predictions, major disasters were interpreted as heralding the end: the sack of Rome, the fall of Jerusalem to Muslim armies, the Black Death, the Thirty Years War. Repeatedly, the last days would fail to materialize, and a new generation would set to work scratching out revised due dates.

This awkward history has tended to give apocalypse an unsavory reputation. But all such decoding projects stem from a basic misunderstanding of Christian scripture. The Book of Revelation itself rejects any attempt to use its words to predict the time of the end: "I will come like a thief!" it quotes Christ as exclaiming. This striking phrase echoes similar disclaimers in the letters of Paul and Peter, as well as Jesus' emphatic statement in the Gospels: "No one knows the day or the hour." Radical uncertainty is ours to live with.

BACK WHEN THE COVID PANDEMIC still felt fresh, it was briefly popular to speak of "these troubled times" – for example at the beginning of business emails, as a signal to the recipient that one was aware that this particular invoice or complaint wasn't really that important, given all the truly bad things that were happening.

Though such phrases quickly became tedious, they point to an attitude it would be wise to cultivate. The times *are* troubled; they almost always have been. Our troubled times are (probably) not the end of the world – but they may be manifestations of the plagues that Revelation describes as being poured out on humanity. That in itself, strangely enough, is reason for hope.

The word *apocalypse* appears in several key places in the New Testament apart from the Book of Revelation. It appears, for example, in Paul's Letter to the Romans, though English speakers often don't realize it when reading a translation. "The creation waits with eager longing for the revealing" – *the apocalypse* – "of the children of God." Paul continues:

The creation itself will be set free from its bondage to decay and will obtain the freedom of the glory of the children of God. We know that the whole creation has been groaning in labor pains until now; and not only the creation, but we ourselves, who have the first fruits of the Spirit, groan inwardly while we wait for adoption, the redemption of our bodies. For in hope we were saved. Now hope that is seen is not hope. For who hopes for what is seen? But if we hope for what we do not see, we wait for it with patience.

These lines describing a cosmic rebirth ("labor pains") abound with a surplus of meaning that has inspired countless pages of commentary. One thing at least is clear: This apocalypse – our own apocalypse, to borrow Paul's language – promises that there is an afterward to the fact of death. While the sufferings of the present time are real enough, the final word on humanity, and on the earth itself, will not be meaningless extinction. Even our bodies have a future, Paul says, though they will be transformed in a way that remains mysterious. The resurrected Jesus – a flesh-and-blood person who in the Gospels eats a meal, breaks bread, and roasts fish at a lakeside campfire – is proof and pioneer of what resurrected humankind will be.

The Talmud tells how one of Paul's contemporaries, Rabbi Yohanan Ben Zakai, used to say: "If you have a sapling in your hand and they tell you 'The Messiah is coming!' first plant the sapling and then go to greet him." In the interim of the ages, as the universe's great Sabbath approaches, humankind has work to do. Plant the sapling; tend the earthworms; welcome the children given to you; hope. The times may be troubled – but beyond them, there's a future to eagerly await. ➤

Radical Hope

When worlds die, we need something sturdier than
the myth of technological and social progress.

PETER J. LEITHART

THE YEAR 2020 CAME DOWN like the
wolf on the fold. Then came 2021. And
2022. It feels like "the end of the world as
we know it." It feels like an apocalypse. It
may be one. Worlds *do* die.

Historians and junior high students debate the
precise end of the Roman Empire and whether it
should be described as a "fall," but no one doubts the
Roman Empire now lies peacefully in the graveyard
of history. Remnants of medieval life persist in our
world, more than we realize, but we no longer live
medievally.

Worlds can disappear speedily. Less than a month
after the storming of the Bastille on July 14, 1789,
France's National Constituent Assembly abolished
feudalism and the mandatory tithe, shattering the
foundations of medieval order and slashing the alliance between the French monarchy and the Catholic
Church that began with Clovis's baptism in the early
sixth century. Within two years, the royal family fled
the palace and early in 1793 Louis XVI was executed.

More recently: the world that existed before the
Russian invasion of Ukraine is gone, a memory of
the age of American unipolarity and what was in
retrospect a shockingly fragile European peace.

The change was rapid and distinct: the week after the invasion, one felt a nostalgia for a stable geopolitical order that simply didn't exist anymore. Once it was destabilized, its former stability in retrospect looks illusory.

Periods of rapid transformation are preceded by years of slower and subtler change. Before revolution broke out, France simmered with decades of intellectual, cultural, and political ferment. Diderot's *Encyclopédie* was published between 1751 and 1772, and few of the *philosophes* lived to see the Bastille fall.

All artwork by Anna and Elena Balbusso

Worlds die. Ours may be dying. But the crises of the past three years aren't isolated.

Decades before the Bolsheviks coalesced into a juggernaut, Dostoyevsky spied intense proto-Lenins among young Russian intellectuals.

Transitional periods can last for decades, even a century. One hundred and twenty years passed between the rise of a Pharaoh who didn't know Joseph and Israel's entry into the land of Canaan. For over a century, Israel was either oppressed by Pharaoh, living in desert tents, or fighting to capture the land of promise.

Since biblical times, plagues and wars have marked the end of an age. As recounted in the Pentateuch, when Pharaoh refused to let Israel go, Yahweh sent a severe pestilence on Egypt's horses, donkeys, camels, herds, and flocks. After the Exodus, Yahweh repeatedly warned Israel not to be complacent; he threatened to afflict the Israelites with the plagues of Egypt if they failed to keep the covenant. Six centuries later, as Judah's monarchy slouched toward exile, the prophet Jeremiah warned of a multiple judgment on the land – sword, famine, and pestilence. Jeremiah's threats reinforce each other: Invading armies kill men and ruin the land, leaving it incapable of supporting the remnant that survives an invasion. Weakened by malnutrition and surrounded by looters, the people are sickened by foreign germs. No wonder sword, famine, and pestilence are three of the four horsemen of the apocalypse (Rev. 6:1–8). When war and pestilence appear, the end is nigh.

Pandemics have played an outsized role in the fall of empires ever since. In the mid-fourteenth century, the Black Death killed between twenty and fifty million, one-third to three-fifths of Europe's population. The Arab historian Ibn Khaldun lamented that this bubonic plague pandemic "swallowed up many of the good things of civilization and wiped them out in the entire inhabited world." More than a century of devastation followed – war, famine, social decay, and disease – which readied Europe for the message of the Reformers. We know the fate of Rome "was played out by emperors and barbarians, senators and generals, soldiers and slaves," yet it was "equally decided by bacteria and viruses, volcanoes and solar cycles," classicist Kyle Harper suggests. Late Roman history could be called "the age of pandemic disease," he writes, with major outbreaks in the second, third, and sixth centuries. The Antonine Plague (AD 165–80) may have killed as many as seven million. By comparison, Rome never lost more than twenty thousand men on a day of battle. For ancient Rome, "germs are far deadlier than Germans."

What happened to the Roman Empire has happened regularly enough for geographer and historian Peter Turchin to propose a

All artwork by Anna and Elena Balbusso c/o Theispot. Used by permission

theory of "secular cycles" of global expansion and contraction. Globalization is a recurring historical phenomenon. The human race "has experienced other periods of heightened long-distance connectivity that resulted in massive long-distance movements of goods, people, ideas, genes, cultivars, and pathogens." During the Age of Discovery, for instance, "all major population centers of the world, both in Afro-Eurasia and the Americas, were connected by trade and conquest." As people increase their contacts across the globe, they're more likely to pass on viruses and germs. The probability of pandemic increases, and pandemics, in turn, are often harbingers of a degenerative cycle. The expansion phase is "relatively disease-free" but "epidemics are much more likely to occur during the stagnation phases of secular cycles." Population growth eventually crosses "the epidemiological threshold (a critical density above which a new disease is able to spread)." Pandemics and social upheavals reinforce each other. Bacteria and viruses flourish in periods of declining living standards, migration, and urbanization. Global contraction correlates with population decline, and Turchin claims that "a pandemic or a major epidemic is frequently (but not always) the primary cause of population decline, and the trigger for the crisis."

Worlds die. Ours may be dying. But the crises of the past three years aren't isolated. They fit into a longer and larger pattern of events that seem to be bringing a world to an end and giving birth to something quite different.

Contraction of the West

Suppose we're in a transitional age. Suppose a world is ending. We still need to ask, *What world is ending?*

First answer: A world controlled by the power and values of Western Europe and North America. In the fifteenth century, Western Europe embarked on a half-millennium of global adventuring, migration, settlement, and colonization. Then, in the mid-twentieth century, it stopped. Europe's former empires *de*colonized themselves in the 1960s; Europe wouldn't think of *re*colonizing. Debates about "American empire" testify to the shift. No one doubted Britain had an empire; many question whether America's global hegemony amounts to "imperialism."

The global economy provides a good measure of the change. Western production, trade, and finance still dominate the globe, but three of the top five economies are Asian. The United Kingdom, France, Italy, and Canada are still in the top ten, but have been joined by Brazil.

In particular, China is leveraging its Western-aided prosperity to carve out its own zone of economic power. Before Russia's invasion of Ukraine, the Chinese government had planned to invest 1.4 trillion dollars to create a twenty-first-century "Silk Road," the "Belt and Road" transportation web that will link Asia to North Africa and the eastern edge of Europe – sixty-five countries and over four billion people. China hopes Western Europe will be lured east. Plus, China produces most of the world's antibiotics and pharmaceutical components, and Chinese nationals own leading American entertainment companies, as well as real estate and many American businesses. In 2019, Daryl Morey, general manager of the NBA's Houston Rockets, tweeted his support for dissenters in Hong Kong. It became an international incident and cost the league hundreds of millions of dollars.

Peter J. Leithart is the president of Theopolis Institute for Biblical, Liturgical, and Cultural Studies in Birmingham, Alabama. This piece is an excerpt from his book God of Hope, *forthcoming from Athanasius Press.*

A year later, Morey quit. Even in basketball, the unipolar world is no more.

The evolution of the church is a further measure of Western contraction. There are still state establishments (in England, for instance), but Western politics and culture haven't operated by Christian norms for a long time. Christian symbols and beliefs no longer provide the fundamental framework for public life, nor for many individuals.

At the same time, a "new Christendom" is taking shape in the Global South. At the time of the Reformation, Christianity was largely confined to a shrinking Europe. Since then, it has expanded to every corner of the globe, becoming the main religion in the Americas, Australia, southern Africa, and Pacific islands. Today, the majority of Christians reside in Africa, Asia, and Latin America.

North American and European Christianity still leads in many ways. Western churches are wealthier, and their influence is buttressed by the considerable geopolitical power of Europe and North America. Western schools educate theologians and leaders from the Global South. Yet, on all these fronts, the tide is turning. Africans have gained considerable clout in the Anglican Communion, often strengthening the position of beleaguered traditionalists in England and North America. Pope Francis is Argentinian, and he's likely the first of many non-European popes. Christianity has ended its sojourn as a "Western" religion, as the world is no longer a Western playground.

Second answer: This geopolitical shift has been accompanied by an epochal ideological shift. Many among the Western intellectual elites have adopted a post-colonial outlook, which views the West as the main source of evil in the world. No reasonable person believes Western civilization is innocent – what civilization is? Perhaps more importantly, few believe that it is admirable. Western thinkers have formulated appallingly racist theories and leaders have committed sickening atrocities; as just one of many examples, Adam Hochschild's 1999 study of colonial Africa, *King Leopold's Ghost*, is a particularly haunting narrative of exploitation in the Belgian Congo. But admitting those evils is different from saying the whole civilization is unjust and racist to its roots, that what is distinctive in it is tainted entirely. No civilization can flourish when its elites begin to assign it to history's overfull dustbin. Societies that have no confidence in their shared principles cannot survive for long.

As the modern West's influence contracts, its post-Enlightenment values also go into retreat. Old-fashioned liberalism of the "I abhor what you say but I will defend to the death your right to say it" variety has died. Progressivism has become the *de facto* established religion of swaths of the United States and other countries, and it is a jealous religion. To evade social and professional repercussions, one quietly censors oneself. It's fruitless to protect Western liberalism, since there is no longer a liberal West to protect.

Western ideals are losing their power to energize *non*-Westerners too. Beginning in the Enlightenment, Western thinkers promised to liberate the human race from the "irrationality" of superstition and religion. If we can't eliminate irrationality entirely, at least we can keep it out of public life, so it doesn't do so much damage. Religion arouses irrational passions; politics should be conducted by reasoned deliberation. Religion is violent; purging it from politics will yield a utopia of nonviolence. Advanced, "Westernized," nations do the right thing and privatize religion.

It was always a ruse. That Empire of Reason is, of all empires, the most thoroughly dust-binned. Religion has never been, *can* never be, eliminated from public life. Western regimes, like all other regimes, have *always* been intertwined with religion: regulating it, supporting it, being supported by it or critiqued by it. But many believed the ruse, including sociologists who were convinced that modernization, industrialization, the expansion of technology and education, and the establishment of democratic regimes would naturally produce secular societies, where religion was a private consolation for a diminishing handful of traditionalists.

Events torpedoed the secularist dream. One of the big stories of the past half century is the re-emergence of political religion. Secular politics received a massive, albeit often unnoticed, refutation in the disintegration of the Soviet bloc during 1989–91. A world order seemed to vanish overnight. Maps were hastily redrawn to include the Muslim-dominated "stans" of Central Asia and the newly independent Balkan states of the former Yugoslavia.

Western leaders saw it as a triumph of Western democracy, and expected Eastern Europe to become democratic and capitalist, and, above all, to stay secular. They were wrong. As the English political philosopher John Gray saw at the time, the Soviet bloc fell to movements driven by religious fervor and nationalist passions. Poland had begun to wobble during Pope John Paul II's 1979 pilgrimage to his homeland. His masses drew millions. Christian processions passed through the streets of Polish cities. The final blow to the Soviet empire didn't come from the West but from a revived *Russian* nationalism. That nationalism, again on the march, is supported by a thoroughly public brand of Orthodox Christianity.

Hope takes risks; it risks disappointment. Hope is too frightening. Acedia evades the pain of hope by hoping for nothing.

What followed the collapse of the Soviet bloc was not a Westernization of Eastern Europe. Gray writes: "In the wake of Soviet communism, we find, not *Homo Sovieticus* or any other rationalist abstraction, but men and women whose identities are constituted by particular attachments and histories – Balts, Ukrainians, Uzbeks, Russians, and so on." The year 1989 began a return to "history's most classical terrain of ethnic and religious conflicts, irredentist claims and secret diplomacies."

We live in an age of political re-enchantment, awakened from the Kantian dream of European perpetual peace, the rationalist project of secularization on as many continents as possible. As the Russian army continues its attacks, we are reminded that not all re-enchantment is good. As all that we took to be solid melts, we feel as if we're trying to walk on water. As the world rocks and roils under our feet, anxiety can drive out hope.

Postmodern Mood

Anxiety isn't new. In the West, the cultural mood has been shading to black for some time. Hope has been in retreat: perhaps ironically, in the relative stability and prosperity that existed before the pandemic and the war, we were already an anxious bunch, as evinced by our taste for dystopian films and novels.

This perpetual anxiety's other side is what medieval thinkers called "acedia." As Paul J. Griffiths writes, acedia

is the mark of those sufficiently habituated to looking at nothing that when they look at something – and most especially at the Lord, the supreme object of delight – they can only sigh, shake their heads, and close their eyes. . . . Long looking at nothing saps the energy and dulls the perceptions so that when sinners are faced by something they lack the energy to respond to it with the joy that all somethings – good and beautiful just because they are something – require of the gaze that sees them for what they are.

Hope takes risks; it risks disappointment. Hope is too frightening. Acedia evades the pain of hope by hoping for nothing. Acedia is the dreary, blank-eyed "whatever" that follows the collapse of our ancestors' dreams of liberation and progress. It's the postmodern mood rising from the smoking ruins of the modern eschatologies that have organized Western life for several centuries.

Of course, there are still buoyant believers about, like Steven Pinker, who cheerily celebrates our ever-improving species. The number of people in extreme poverty has been falling for decades, the majority of the planet is free of war, child and maternal mortality are down, illiteracy is down. Progress will continue indefinitely, Pinker thinks, so long as we stick to our Enlightenment guns and follow the science. This line of thinking is taken to the extreme by the technophile transhumanists, who dangle the promise of liberation from all human limits, including, perhaps, the limits of death.

Behind this apparently hopeful project is a thinly disguised despair. Techno-utopianism's hope is contemptuous of human beings as they actually are. Its future belongs not to humans, but to burnished, glossy super-humans.

By and large, we've lost confidence in our myths. Even our Promethean hubris is a cloak for our Sisyphean despair. Our feet find no solid place to stand. We reach out for something firm to hang on to, but our hands grasp at the air. "We come to an end without hope" (Job 7:6).

Designed for Despair

The despair is predictable. Since the Enlightenment, the West has devoted itself to erecting a world that can flourish without God. As Paul said of the Gentiles, we're "without God" and for that reason "without hope in the world" (Eph. 2:12). Modern Western civilization has entered a cul-de-sac, the dead end of our own achievements. To a certain degree, what we hoped for has come to pass. We are living in the dreamed-of, hoped-for future. It is not good enough. It has made our lives, in many ways, less human.

Think, for instance, of our addiction to acceleration. Our communications are instantaneous, our travel more rapid than at any other time in history. Social change accelerates, as each generation develops its own culture. Yet despite timesaving devices, we feel we have less time than ever. We eat faster, sleep less. Our conversations are short, as are our attention spans. We lack the leisure that allows us to be human.

Many of our political practices run on a different clock from technology, and this desynchronization dislocates social life. Western states, for instance, insist on a participatory form of government. But democracy takes time. Frenetically trying to keep up with our technologized pace, we do not take time to listen to one another. We don't have enough time to get to the bottom of our disagreements, much less to resolve them.

This is, if anything, even more true of the process of diplomacy. Traditional diplomacy is often a matter of a deliberate slowing down: we stop, we listen, we halt a rush to war. We allow time for procedures and discussions, held in secret, to avert a cascade of half-considered escalatory decisions. Slowness and professionalization are the essence of diplomacy. Traditional diplomacy is the antithesis of Twitter-based populist foreign policy dictated by the mob.

Technological acceleration clashes with modernity's promise of freedom. This acceleration is closely linked to addiction, and it is a paradox. On the one hand, we have far more freedom of choice than past civilizations; on the other, we have less freedom to opt out of that very technology. We are economically and in many cases personally addicted to the ease of technology, to its dopamine hits. We are less skilled, and thereby less free. We are trapped, and we know it.

The imperatives of the Enlightenment project require constant improvement, and refuse us the resources of the past. We must question tradition. Nothing is accepted on authority, because authority comes from outside, not from within. We long to be free from dependence on and submission to another.

Yet despite its apparent optimism, this bid for autonomy relentlessly kills hope. As the Jesuit psychologist William Lynch observed, hope is necessarily mutual. Hope correlates to *help*. I breathe, and implicitly hope the world will respond to keep me alive. I work, hoping to achieve something. I love, hoping for a lover to return my love. Being well means being hopeful: "The well hope for a response from the world." Help comes from *outside*: from God, others, the world. Our need for outside help "is deeply inscribed in every part of us and is identical with our human nature." The pursuit of autonomy, the denial of dependency, cruelly reinforces the despair of those not able to fake it well, those who can't disguise their fundamental neediness. The idol of autonomy suggests that their neediness, their inability to do it all themselves and to reinvent themselves, renders them subhuman. It suggests that the helpless can't be members of the human race until they've learned to help themselves.

Modernity's very success attenuates hope. We are trapped in an arid present: it is unfashionable to believe that we have something of value to pass on to our descendants; it's equally unfashionable to believe our ancestors have anything of worth to give us. As Alastair Roberts remarks, "With the improvement of our life conditions and the saturation of our horizons with earthly pleasures and diversions, Christian hope gets squeezed out and our relationship to death changes." Because of our extreme mobility, our world isn't hospitable to the development of immaterial social and spiritual goods, or even to physical ones that are expected to last beyond our time. As a result, "people are unlikely to make sacrifices that depend upon future generations for their payoff."

Perhaps most fundamentally, our Godless world is a story-less world. "Modernity was defined by the attempt to live in a universal story without a universal storyteller," writes Robert Jenson. It cannot be sustained. "If there is no God," Jenson concludes, "there is no narratable world." And without a narratable world, without the apocalypse as its triumphant and awe-inspiring denouement, there can be no hope.

Radical Hope

Once Christian hope, hope in a kingdom and an age to come, infused the Western world. Cathedrals stretched to heaven, embodying in stone the hope for a helper from beyond this world. Painters depicted Jesus rising from the dead, or enthroned in glory at the Last Judgment, pointing viewers toward the final things. Preachers preached the scriptures given, in Paul's phrase, "that we might have hope" (Rom. 15:4). When the West turned its back on God, it continued to run, for a time, on the fumes of this hope: an apocalypse of technology, of perpetual peace, was always just around the corner. But because this hope trusted in idols, it withered.

What happens when the taproot of hope withers? We need hope to live virtuously, to act with courage and patience; we need hope to act *at all*. To survive, people must find another root of hope. They must locate the sources of

We need hope to live virtuously, to act with courage and patience; we need hope to act *at all*. To survive, people must find another root of hope.

what Jonathan Lear calls "radical hope," a hope "directed toward a future good that transcends the current ability to understand what it is."

Over the two millennia since the birth of Christianity, many worlds have ended, just as our world may be ending now. At such times, it is the task of Christians to nourish hope within societies whose transient hopes have withered. Churches must become communities that cultivate radical hope.

How do we go about this? There's no trick, nor is there any special "ministry of hope." The church *is* a community of hope, and all of the church's ministries and activities express and nourish hope. The word nourishes hope; prayer nourishes hope; singing nourishes hope; baptism nourishes hope; the Lord's Supper nourishes hope. When we open our homes to the homeless, feed the hungry, and clothe the naked, we act in hope and bolster hope, as the Spirit builds our confidence in God's promises and good gifts.

The church's existence, activities, and ministries nourish hope because they are specific avenues of communion with God. God speaks in his word, hears our prayers and songs, claims us in baptism, feeds and feasts with us at the table, shines through us as we go out as lights in the world. God is the God of hope, not merely a God who gives hope or who is the object of hope.

How do churches nourish hope in an age when worlds are ending? By staying close to Jesus, our hope of glory. Simple as that. ⬎

Odilon Redon,
*Bouquet of
Flowers*, 1912

Stopping By with Flowers

I used to bring back flowers for my mother.
I'd stop the car and gather a small bunch.
She'd always be surprised, and always grateful.
She'd put them in a vase. *Could we have lunch?*
I wasn't free, but that part did not matter
so much, I told myself. It was the thought.
She loved my stopping by for those few minutes.
Still, I'd feel guilty, since I felt I ought
to visit far more often, and for longer.
She never said it, but I knew she knew
that I *could* make the time. I'd sometimes linger,
but then I'd go do what I had to do,
hoping that what I could and did not say
might be made up for by that small bouquet.

BRUCE BENNETT

The Sermon of the Wolf

The Anglo-Saxon world was collapsing amid
Viking terror and political chaos. One bishop
held a kingdom together.

ELEANOR PARKER

"Beloved ones, know what is true: this world is in haste, and it is nearing the end." —Wulfstan, archbishop of York, AD 1014

THESE APOCALYPTIC SENTIMENTS could have appeared yesterday on the front page of the *New York Times*, or two thousand years ago in an epistle to the early church. In fact, they were written just over one thousand years ago, by an Anglo-Saxon bishop. They are the opening of a sermon preached by Wulfstan, archbishop of York, in AD 1014. He begins by warning his hearers that the state of the world is getting worse, and that this can mean only one thing: "In this world the longer things go on, the worse they are, and so it has to be that things grow very much worse because of people's sins, before the coming of Antichrist, and then indeed it will be grim and terrible throughout the world."

Like many medieval Christians, Wulfstan believed that the end of the world was a prospect with which all thinking people ought to reckon seriously. It seemed only wise: Christ warned his disciples to read the signs of the times, to recognize "wars and rumors of wars," upheavals in the natural world, and the prevalence of sin and violence as indications that the apocalypse would soon be at hand. The end of the first millennium, too, seemed to accelerate the prospect of the apocalypse to Wulfstan's contemporaries, and those who can remember the year 2000 will be well aware how significant dates like the turn of a millennium can produce disquiet, even in a supposedly rational age.

In 1014, however, Wulfstan had specific reasons for feeling that the end times could not be far away. England was in a dire state. Since the last decades of the tenth century,

the Vikings – Anglo-Saxon England's old enemies – had been conducting campaigns of persistent and devastating raids, ever growing in intensity. The pressure to resist them had drained the country financially, militarily, psychologically, emotionally. King Æthelred and his advisers tried various strategies to combat the danger, but the scale of the challenge was tearing them apart; the court had descended into factionalism, and the king himself was increasingly blamed for all that was going wrong. There's a reason that Æthelred has become known to generations of schoolchildren as "Æthelred the Unready," a play on words which implies not lack of preparation (though he could be accused of that) but the idea that the king was badly counseled or ill-advised. By 1014, England was facing not just piracy and internal disputes but the real threat of invasion. At the end of 1013, Æthelred had been forced to flee England and the country had fallen under the full command of the Danes. Barely a month later, the sudden death of the Danish king Sweyn unexpectedly turned things upside down again, and Æthelred was able to come back. However, Æthelred was only invited back by his advisers, according to a contemporary source, on the condition that "he would govern more justly than he had done before" – a phrase which seems to suggest little faith that he would do so.

In his apocalyptic sermon, Wulfstan speaks to a country in total moral collapse. "Nothing has prospered now for a long time, at home or abroad," he says, listing "harrying and hunger,

Opposite: Figurehead on replica Viking ship outside the Vancouver Maritime Museum

---

Eleanor Parker teaches medieval literature at Brasenose College, Oxford, and is the author of Dragon Lords: The History and Legends of Viking England *(2018) and* Conquered: The Last Children of Anglo-Saxon England *(2022).*

Opposite:
Viking
attack,
from a
Saint-Aubin
manuscript,
ca. 1100

burning and bloodshed," plague, pillaging, and unfair taxation among the many evils afflicting the nation. But it isn't a warmongering sermon; in fact, Wulfstan says very little about the enemy. He's much more concerned with the idea that under sustained and unrelenting pressure, the English people have abandoned their obligations to family, community, and country. "The devil has now led this nation too much astray for many years, and there has been little loyalty among men, though they spoke well, and too many injustices have reigned in the land," he tells his hearers bluntly. "Daily one evil has been piled upon another and injustices and many violations of law committed all too widely throughout this entire nation."

Wulfstan goes on to detail all the different ways in which the essential bonds holding society together had begun to break down. He gives specific examples, described in a manner intended to horrify and disgust his audience. His account still has the power to shock a thousand years later: particularly horrible are his descriptions of desperate people selling their own family members into slavery and his fierce denunciation of the sexual abuse of female slaves. There is no fudging of the details, no opportunity for the audience to turn their eyes away; in stark and brutal language, Wulfstan makes his audience confront exactly what their society has become. No wonder he thought the coming of the Antichrist could not be far away.

This sermon is known as the "Sermo Lupi," the "Sermon of the Wolf." The title plays on Wulfstan's own name, but it is also an apt characterization of the angry tone of his urgent message. In Anglo-Saxon culture, the wolf was the ultimate outsider: the archetypal outlaw, a dweller in the wilderness and on the borderlands. This title suggests that Wulfstan, at least on this occasion, embraced the wolf as a facet of his identity: it is not gentle preaching but the howl of a wild beast, coming out of the

darkness. It's meant to frighten you, to send a shiver down your spine.

But Wulfstan was no prophetic voice crying in the wilderness; he spoke from the heart of power and influence. He was an active and experienced politician who had served first as bishop of London, then as bishop of Worcester and archbishop of York, making him the country's second most senior church leader by 1014. He was an adviser to King Æthelred and had been involved in contemporary debates about how to respond to the threat of the Vikings – whether by fighting them, paying them off, or trying to convert them, all strategies used at different moments, with different levels of success. In these times, though, power was no guarantee of personal safety. In 1012 Wulfstan's senior colleague, the head of the English church, Ælfheah, archbishop of Canterbury, had been murdered by a Viking army. Whether the end of the world was coming or not, Wulfstan might have felt there was a good chance his own end wasn't far away. Death had become a daily reality.

We can only guess what personal fears and griefs lay behind the howl of anger in his sermon. But – wolf though he was – Wulfstan was also the very opposite of an outlaw; he was a maker and writer of laws, who had worked on the formation of Æthelred's legal codes. Law was central to his thinking, and it's at the core of his sermon. Many of the terrible sins Wulfstan describes are diagnosed by him as breaches of law, a concept which for him had the broadest possible meaning: it involved not only actual crime, but violations of personal integrity, such as breaking your word, neglecting your duty to your parents, or disregarding your marriage vows. It also covered the breach of religious obligations like fasting and the mistreatment of church property.

We wouldn't think of all these as questions of law today, but for Wulfstan they were all

part of the same thing: God's laws, the laws by which society governed itself, and each individual's moral commitment to his or her obligations should all be connected. They involve the important value of what he calls *getreowþ*, an Old English word related to our word "truth," which encompasses loyalty, honesty, trustworthiness, and integrity.

As he saw it, all these were in jeopardy in 1014. However, for Wulfstan diagnosing his society's ills as breaches of law was not a source of despair, but an opportunity. It meant he could offer a plan of action. In this sermon his purpose is not just to denounce and lament, to criticize without providing solutions. His aim is to preach repentance and amendment – to convince people that things can get better, even in the shadow of the end times. The end will come; he has no doubt of that, and right now things are almost as bad as they can be. But there are measures we can take in the meantime, he suggests, things that will help. They won't stave off the apocalypse or keep the Antichrist away. Yet they're still worth doing – both morally right in themselves and a remedy for present evils.

His message is simple: repent, repair, do better. There's no pretense that it'll be easy. "A great wound needs a great remedy," he says, "and a great fire needs a great amount of water if the blaze is to be quenched." The worse the situation, the more work and collective effort it

will take to mend it. But the promise that it can be mended is, nonetheless, a remarkably hopeful takeaway from such a fierce and angry sermon. As a response to the threat of coming apocalypse, it's almost optimistic. Wulfstan asserts a firm belief that doing what is right doesn't cease to matter, even if time is running out. In fact, it matters more. Living in the shadow of the end of the world doesn't involve giving up on life, but recommitting yourself to what is most precious about it. For Wulfstan, that means especially strengthening the bonds we share with other people – our families and communities, the vulnerable and the poor, those we trust, and those who need to be able to trust us.

Though directly confronting the problems of his own time, Wulfstan's sermon speaks powerfully to our time as well. His picture of a society suffering from a breakdown of "truth," in which both institutions and individuals have succumbed to corruption and self-interest, is uncomfortably familiar. So too is his warning that in the face of sustained crisis, a society can lose sight of what it values and believes to be true; fear and despair can make people do terrible things, sacrificing even what they hold most dear.

There is an emotional resonance in Wulfstan's opening statement that the world is "in haste," advancing towards its end at a frightening speed. Today, people often talk

Fourteenth century illumination depicting kings Edmund Ironside and Cnut, from the *Chronica Majora*

about feeling that the pace of life is speeding up, and we tend to attribute it to modern technology and rapid social change. Over the past two years especially, between political crises and pandemic and war, people have joked wryly that 2020–22 has felt like a decade or a century all rolled together. The problem is not really the speed at which things are happening, but the feeling that we can't keep up with them. We struggle with the disparity between the pace at which events are unfolding around us and our ability to process them, mentally and emotionally – barely finished with one once-in-a-generation crisis, before we're suddenly plunged into the next. It's like being caught in a fast-flowing river, swept along by the stream, just trying to keep one's head above water.

It seems the people of Wulfstan's time felt the same. In their time and our own, this sense of haste can lead to feelings of helplessness or despair. If things move so fast, and they're only going in one direction, what's the point of trying to change what's happening? But Wulfstan's conclusion stresses action, the belief that there are still things worth doing and steps that we all can take. "Let us do what we need to do: turn toward the right and abandon wrongdoing, and earnestly atone for what we previously did wrong. And let us love God and follow God's laws . . . and have loyalty between us without deceit."

Wulfstan practiced what he preached. In 1016, after two more years of devastating warfare, the English king was finally defeated and the Vikings were victorious. England was conquered, and the Danish prince Cnut – a young Viking whose actions so far had shown evidence of nothing but brutal cruelty to his enemies – became king. Wulfstan might have thrown up his hands in despair, thinking all his warnings had been in vain. What advice could he offer to a king like this? What hope could he preach now?

But he didn't give up. He still had work to do. All along, his priority had been the moral health of the nation, and he still believed that the unifying power of law could offer a cure. So Wulfstan became Cnut's chief English adviser, writer of the king's laws and public pronouncements, which emphasized the need for conquerors and conquered to live in harmony, to observe the same laws, to share common values, and to find reconciliation after the long years of war. This time, Wulfstan's work had real success. Cnut proved to be susceptible to influence; peace came, at least for a time, and the laws they made formed the basis for many later codes, ties that still sought to hold English society together centuries after Wulfstan himself was dead.

Perhaps "this world is in haste, and it is nearing the end," but it always has been. Whether it comes soon or late, the end is always coming. We don't know what will precipitate the end times; it turned out not to be the Vikings, or the year 1000, but it might yet be nuclear war or climate apocalypse. If we can learn anything from those who lived under the shadow a thousand years ago, it is that there is work worth doing even in the darkest times. Wulfstan found his hope in restoring the bonds of community and family, faith and justice – bonds forged not only by rulers and social reformers but by individuals and their choices, however small each of these may seem. Whatever the darkness of the times we live in, some good can yet be done by every turn toward the truth. ➘

> "Let us do what we need to do: turn towards the right and abandon wrongdoing."
> —Wulfstan

The New Malthusians

LYMAN STONE

Population pessimists claim that having children threatens the environment. They are wrong.

Previous spread: db Waterman, *It Seemed Such a Good Idea*, collage and acrylic on paper, 2019

I N 1824, a young American doctor, freethinker, and atheist named Charles Knowlton was imprisoned for digging up bodies and dissecting them for medical research. This experience served to encourage rather than deter him in his efforts. By 1832, he had published a book called *Fruits of Philosophy*, which acted as a reference guide for his patients in rural Massachusetts, with suggested treatments for ailments ranging from infertility to impotence to unwanted pregnancy. Modern scientific understandings suggest that its recommended contraception methods were unlikely to have been effective, but the key point is that they gave people clear guidance saying that they *could* control their own fertility. The book was declared obscene, and Knowlton did another stint in jail for the crime of distributing it. But from 1800 to 1850, birth rates in Massachusetts fell from about 5.4 children per woman to about 3.3.

Across the Atlantic, other movements were brewing. The Anglican cleric and noted economist Thomas Malthus had published the first edition of his *An Essay on the Principle of Population* in 1798. It received excellent reviews, but in those pre-internet days, it took some time for its ideas to spread. It was ultimately the greatly expanded sixth edition, published in 1826, which presented Malthus's full argument that population growth would inevitably exhaust the physical resources of the planet, leading to misery, vice, and economic depression. This edition would be read and cited enthusiastically for generations to come, influencing the thinking of Charles Darwin and other early biologists.

But when Malthus spoke of "vice" as one of the possible forces restraining human population growth, he had a quite particular view of what he meant. Certainly he thought of war and murder as population limiters, but these he more generally categorized as "misery." The specific vice Malthus had in mind as population-limiter was sexual licentiousness. That is to say, his worry was not only that humans would starve to death, but that, if we succeeded in avoiding starvation, it would be through *vice* (that is, birth control, abortion, and the spread of venereal disease) rather than *virtue* (abstinence). Specifically, he says: "Promiscuous intercourse, unnatural passions, violations of the marriage bed, and improper arts to conceal the consequences of irregular connexions, clearly come under the head of vice."

Thus, for Malthus, one of his great *worries* was that overpopulation would lead people to use contraception and have lots of non-procreative sex. Far from being co-conspirators, Malthus and Knowlton could hardly have been more opposed to one another, philosophically speaking. Yet from these two quite different strands – a body-snatching Massachusetts atheist and enthusiast for birth control alongside a British economist and Anglican priest worried about excessive sexual indecency – was birthed a movement. That movement, curiously enough, has taken the name of the man who would have been most uncomfortable with it: Malthusianism.

MALTHUSIANISM TOOK some time to coalesce into a meaningful ideology. Certainly Malthus's economic ideas informed the terrible British response to the Great Hunger in Ireland, worsening the famine there (Malthus favored the restrictive

---

Lyman Stone is the director of research for the consulting firm Demographic Intelligence, a research fellow at the Institute for Family Studies, and a PhD student at McGill University. He lives in Montreal with his wife and their two daughters.

"Corn Laws" which intensified the starvation conditions). Beyond that, his writings were influential in establishing the regular British census system from 1801 onwards, for the express purpose of tracking population growth. But a systematic ideology of concern for overpopulation was unimaginable in early-nineteenth-century England, where there were few practical means of preventing population growth.

No country in Europe had any sustained decline in fertility between 1800 and 1870 except for France, where the cultural changes initiated by the French Revolution led to lower birth rates. But by the 1870s, Malthusianism's time had come. Malthus and Knowlton were both long dead, but their ideas, and especially their books, lived on. The spread of ideas about natural selection and evolution had changed public attitudes toward the question of population; it was becoming acceptable to talk, quite in the abstract of course, about who should have babies and who shouldn't. By the 1850s and 1860s, England had a robust, secular civil society, challenging the preeminence of organized religion.

In 1876, two British secularists, Charles Bradlaugh and Annie Besant, republished Knowlton's *Fruits of Philosophy* for a British audience. They were immediately slapped down with censorship and anti-obscenity laws. As the courtroom drama dragged on, it was covered widely in the popular press, attracting enormous readership and public attention throughout the British Empire. And wherever newspapers covered the court case, sales of *Fruits of Philosophy* skyrocketed, and fertility rates plummeted. Despite offering little direct scientific information about birth control, somehow the Bradlaugh-Besant trial succeeded in reducing fertility.

Recent academic research has confirmed that this association is more than coincidental:

exposure to media coverage of the Bradlaugh-Besant trial probably did lower fertility. English districts where newspapers covered the case more heavily saw faster fertility declines after 1877. Anglophone settlers in Canada and South Africa saw simultaneous fertility declines not shared by their Francophone or Dutch-speaking neighbors. Recent British immigrants to America saw sharp fertility declines too, as did Australia. Wherever the Bradlaugh-Besant trial made news, fertility fell.

The driving force of this decline wasn't the direct influence of Knowlton's book or the content of the trial. It was about the cultural

db Waterman, *Freestyler,* collage and acrylic on paper, 2021

signal: it's OK to avoid getting pregnant! It's OK to try to avoid conception! Practically overnight, the old orthodoxy had been overthrown, and everybody was talking about this new thing: *birth limitation*. (Presumably people relied on behaviors that, while not foolproof, can often

db Waterman, *December Rain*, collage, acrylic, and drawing on paper, 2019

help avoid conception.) Meanwhile, Bradlaugh and Besant launched a new social-interest group to carry on the work they began by publishing *Fruits of Philosophy*. They gave their group a fateful name: the Malthusian League.

FROM 1877 TO 1930, fertility rates plummeted across Europe and in European-settled places like Australia and the United States. From about seven children per woman after the War of 1812, fertility rates in America had fallen to just 2.2 children per woman by the 1930s. In the longer-settled parts of the United States like Massachusetts, fertility rates fell below the replacement rate of around 2.1 children per woman.

This decline was mostly driven by improvements in education, a shift away from agriculture, increased urbanization, and reductions in child mortality. Malthusian ideology cannot have caused it all. But the exact timing of the decline was, in many places, set off by specific cultural forces, including Malthusian fearmongering over population. And by the 1920s, the Malthusian League had partnered with Marie Stopes to open a permanent family planning clinic in London – the first such clinic in the world.

But then a strange thing happened. In the late 1930s and early 1940s, fertility rates rose around the world. Researchers continue to debate what caused the Baby Boom, but the most likely explanation is a mixture of economic recovery from the Great Depression alongside wartime deployments and dislocations. Throughout the developed world, in countries that had previously been at or near "Malthusian" fertility rates, births spiked. Meanwhile, death rates in developing countries were beginning to decline, and the population of the Global South began to rise. While colonial governments made inhumane efforts to contain local populations, they ultimately failed. Newly independent states adopted a range of policies, some pronatal, some not, but by the 1970s, most of the world's developing countries had adopted explicitly antinatal stances. Indeed, globally, a consensus seemed to have formed: there were too many humans. It was going to cause a disaster. Human population was a ticking time bomb. Malthus's predictions may have been wrong in the nineteenth century, but they were due to come true in the twentieth.

The most famous of these dire forecasts was Paul Ehrlich's *Population Bomb*, published in 1968. Building on the old Malthusian tropes, the book suggested that ongoing population growth would cause widespread famine within a few decades. Instead, thanks to scientific and agricultural innovation, famine has been in decline for decades, even as the global population has more than doubled.

But while such forecasts have proved wrong from one century to the next, the idea persists that individual human lives should not be created because the earth cannot bear them. Today, fears about climate change and the role overpopulation may play in driving future ecological catastrophe motivates some people to avoid having kids. A recent survey conducted by the *New York Times* found that a third of fertility-age women say climate-change fears are among the reasons why they have not had children yet. Meanwhile, groups like Birthstrike encourage women to take what might be called the "Lysistrata option": boycott having a child until climate policy improves.

These movements are, by and large, just a modern incarnation of the same old Malthusianism. They are not a new cultural force, and there is no reason to be concerned that they represent any new pessimism about human life.

But there is a newer, and more concerning, kind of Malthusianism as well. For example, in a 2019 livestream to her followers, Representative Alexandra Ocasio-Cortez asked rhetorically whether it would be responsible for people to have children, given how climate change would alter those children's lives. This argument may sound similar to conventional Malthusianism, but it's actually radically different: it does not argue that overpopulation is bad *for society*, but that the future world will be so bad that the life of a child born today *may not be worth living*. Western society has

---

A global consensus seemed to have formed: there were too many humans. It was going to cause a disaster. Human population was a ticking time bomb.

typically had strong taboos against suicide, and tends to value optimism and hopefulness, and thus this kind of nihilism has tended to be culturally rare. But it is becoming more common. For example, antinatalist philosopher David Benatar has been profiled in the *New Yorker* and has given his take – the "case for not being born" – in many other publications as well.

These arguments are reaching the public. In five surveys I conducted of American women ages eighteen to forty-four totaling over eight thousand respondents, I find that about one in twenty agree that "it would be better for most people if they had never been born." Whenever I write about fertility, someone inevitably responds on Twitter with some variation of, "But do you realize how bad the world is? Who would want to be born into this?" The source of the badness varies. Sometimes it is climate change. Sometimes it is Republicans. Sometimes it's immigrants and sometimes it's socialists. People of all stripes have their reasons for why it would be better never to be born, but the striking thing is how free more and more Americans seem to feel to express that life is fundamentally bad, that on the grand scale of being, nonexistence is better than existence. This is a very real cultural change; it shows movement toward our becoming a society of despair.

RESPONDING ADEQUATELY to antinatalist despair can be challenging. Conservatives tried to silence the Malthusians and in so doing only gave them even more publicity. But while the effect of the old Malthusianism was to enable people to bring down their fertility to more nearly approximate the desire most

---

The striking thing is how free Americans seem to feel in expressing that life is fundamentally bad, that on the grand scale of being, nonexistence is better than existence.

---

people have for two or three children, the new Malthusianism is at odds with widely shared family goals.

A wide variety of surveys, with many different question structures, has repeatedly shown that in the United States, and indeed in virtually the entire rest of the developed world, most women want two or three children – about one more than they are actually having, on average. Antinatalist scolds are pushing an ideology and a lifestyle that not very many actually want. But the upshot is that they drain

societal support for pronatalist solutions that would encourage people to attain their desired family size. It takes only a few very loud and active objectors to poison the well against the kinds of policy change that could make family life more attainable for more people.

More to the point, the basic premise of the new Malthusianism is just as wrong as the old one's. A child born today has no reason to expect a life of apocalyptic dystopia. While forecasts vary, the most pessimistic scenarios I could find, such as a 2021 report by the Swiss Re Institute, suggest an economic contraction of about 20 percent. That could be a significant loss to quality of life, but would only leave humanity as poor as in the late 1990s. More centrist models suggest economic growth may continue throughout the twenty-first century, even with significant warming. There will be problems, but life will go on, provided we want it to.

And yet something is missing from this rebuttal to the new Malthusians. Arguments about fertility preferences and GDP are dry and technocratic; to get to the heart of what is *wrong*, not simply *erroneous*, in the antinatalist view, we have to look deeper. We must ask a more serious question.

If a child born into a world with 20 percent lower income will lead a life that is *not worth living*, what must be said of the poor alive today, or at any point in history? Should those people not have been born? Are their lives of inferior value? A small but significant share of people answer these questions by saying: *yes,* the lives of the poor are not worth living. This brings us back around to the bad old Malthusianism, the population-control movement with its racist and eugenic applications. The only life not worth living is the life of this idea, which continues to resurface throughout history and deserves to be put down once and for all.

Indeed, while hardships are always greatest for the poorest, population control, as ever, lays its hand most heavily on the weak: it is Uighurs in China facing genocide today, not the Han. In Peru, Quechua-speakers were subjected to forced sterilizations, not the urban elite. In the United States, Puerto Ricans have been targeted for antinatal intervention within recent memory. Around the world, poorer women report desiring the most children, and thus any effort at population control inevitably involves making the most dramatic, and coercive, interventions in the lives of those with the smallest political voice. The problem of poverty is not that the poor are too multitudinous, but that the multitudes are kept poor, not least by systems of political inequality. The idea that birth limitation will alleviate the lives of the poor is yet another bootstrap myth: a Yemeni woman who has one child fewer still lives in a country torn apart by war, without access to education, and without enough food. The world facing her may be fearsome, but its greatest problems have little to do with the headcount.

Yes, humanity is broken, suffering, and destructive – yet it is worth carrying on. Humanity is indeed flawed; it never quite gets things right; it creates new problems for itself all the time – yet it is worth preserving. We can do this with any number of policy changes. But the main way we ensure that humanity endures is by having children. We can determine to pass on the light of life as it was passed to us.

db Waterman, *Bright II*, mixed media, 2019

We can respond with a clear confession in deeds as well as words: human life is worth it.

This is not to say that everyone must have some given number of children, or any at all. There are any number of medical or situational reasons why people may forgo childbearing.

There have been bad economies before, and yet through the Dark Ages, the lights of Christendom did not go out, and in peasants' homes and country churches, the catechesis of life conquering death continued.

My point is simply that the argument made by today's new Malthusians, that life is at the point of becoming unbearable, is factually and morally wrong, even as it becomes increasingly prevalent.

Often it is couched in climate terms. But often it is explained in others: the culture has become too hostile, politics too intractable, or the economy too unfavorable to families. The arguments vary, but ultimately, the response is the same. It is the response of Qoheleth to all those who despair: there is nothing new under the sun; enjoy your spouse, your children, and the world as it is. Yes, it is fleeting: all things given us are so.

There have been hostile cultures before. The Hebrews in Egypt defied their oppressors, and by fertility became too numerous to control. There have been intractable politics before, and yet the Jews in the Babylonian Captivity built homes, got married, and started families in an expression of faith that God would keep his promises. There have been bad economies before, and yet through the Dark Ages, the lights of Christendom did not go out, and in innumerable peasants' homes and country churches of Europe, the catechesis of life conquering death continued. Their example is the counter to despair.

Perhaps the definitive modern Christian writing on despair, and thus on a Christian's reasons for hopefulness, is Søren Kierkegaard's *The Sickness unto Death* (1849), a book-length meditation on that occasion where, confronted with the serious disease of his friend Lazarus, Jesus said, "This sickness is not unto death, but for the glory of God, that the Son of God might be glorified thereby." Kierkegaard's preface suggests that Christian teaching "must bear some resemblance to the address which a physician makes beside the sick-bed." He goes on to say that the true sickness of which Jesus was speaking was not Lazarus' illness, but the despair that afflicts so many human hearts, and, in that instance, the despair of those who had no faith that Jesus could raise Lazarus from the dead. This hopelessness, says Kierkegaard, is truly the sickness unto death: to abandon the hope of life, because you believe that things simply cannot get better. Faith, the antidote, asserts – sometimes with reason but sometimes without – that it can be done: *life can be lived.*

The difficulties that face a child born in 2022 will not be trivial. In addition to climate change, shifting geopolitics, technological change, and demographic shifts, to name just a few, there will be difficulties we cannot foresee or imagine. Children born today are certain to drink from a cup of suffering in some way, as have all others before them. But against these challenges, there is the hope of life itself. ➤

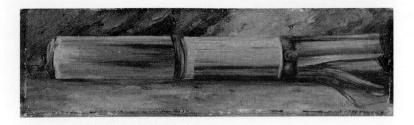

Sugarcane Memories

When I ask him to cut *as'ab*
he hesitates, then shudders it gone.
He pulls a stalk away, shows his grandson
how to stomp it down, crack the stalk at the base
so the roots keep growing, shows him oud,
musical knuckle of root-band, bud furrow and leaf scar.

Canines tear and molars grind, we tongue
the sweet sting, spit pulp and pluck string while he
tells us how neighborhood kids spun quarters and
whacked them with a stick of *as'ab*.
Whoever stuck the spinning quarter got to keep it.
The quarter or the cane I ask, and he laughs.

Then I grew up. He holds the *as'ab* along one forearm
like an offered prayer, drives the knife down,
peels down the purple-green stalk and spins
one about fellahin who were sucking sugarcane
telling stories all night, couldn't find their canes
and had to hobble home in the morning and he laughs
and we laugh hard, as hard as the year.

When I ask if I can help he shakes his head
because he once drove a knife down
into his forearm cutting *as'ab*
and his memory of sweetness with pain is long
and long, as long as the sugarcane.

SHERRY SHENODA

ANIKA T. PRATHER

The Griefs of Childhood

During the pandemic, I learned to weep with my children.

Oswaldo
Guayasamín,
Madre y niño,
ca. 1985

RAISING CHILDREN has always been hard, but the past two years have been especially challenging. All parents desire to have their children grow up in a time of joy and memory-making. However, for our family the very first week of the pandemic threw us into a season of shocking deaths of close family and friends, and I had to find some way to parent while grief threatened to consume me.

I am not sure why this time of Covid also seemed to unleash other types of tragedies, but that's what happened in the midst of missing so many of those we dearly loved. There was the fear of catching Covid. There was the fear of racial unrest and injustice finding its way into our peaceful family life. There was also the fear that comes from wars breaking out around the world. My children felt all of this tension, and I wrestled with how to help them cope while also working through my own anxieties.

I was not sure if I was supposed to put on a strong face for my children and then cry in privacy and silence when they weren't looking, or whether I should let them see me wrestle through my emotions to show that it is OK to feel grief. I do not have a deep reason for why I chose to let them see me grieve. I simply could not do otherwise; I was too overwhelmed. My mind was so full of hurt and fear that I did not have the strength to put on a brave face. To my surprise, as my children and I grieved together, I saw them grow and mature in powerful ways,

Anika T. Prather is the founder of The Living Water School and is a full-time lecturer at Howard University. She and her husband Damon have three young children and live near Washington, DC.

developing an understanding of the Christian faith that I pray will carry them through the rest of their lives.

The author with her family

I RUN A SMALL CHRISTIAN SCHOOL called The Living Water School that I founded in 2015. Before the virus we had been a regular school community with a building and a hectic schedule of classes, sports, plays, assemblies, and more. In March 2020 that all came to a screeching halt. Along with everyone else, we closed down the school and went to virtual learning. But as my children and I walked back into our house, leaving behind a building that had been our second home for five years, we did not feel sad. Instead, we were excited to be home together and we had plans of enjoying our solitude as a family, while also juggling online learning. We did not know what was to come.

What we thought would be a peaceful stretch of family time turned tragic within days. The following Friday morning, we got the news that a very close friend had been brutally murdered. He had grown up in our church and we had just seen him speak there the other day; it didn't seem like it could be real. Then within hours, we learned that my brother-cousin (as our family calls cousins who feel more like siblings) died of Covid after being sick for only two days.

Both those blows in one single day left me falling on the floor in utter grief and despair. Through the blur of my tears I could see my children's puzzled faces, tears streaming down their cheeks too. This cousin, more like an uncle to them, had been their Sunday School teacher and had the most precious way of interacting with our kids. Instantly, he was gone. I gathered my children around me in my arms and we cried together. We talked about our memories of him together. We questioned God together. We expressed fear together. We began to process the loss together. In that moment, as I invited my children into my grief, they invited me into their grief in turn, and the beauty of this is that we never felt alone.

My husband was the strong one. He grieved too, but he figured out ways to bring us joy in such dark times. He got us out into nature and planned all types of family activities to help us take our minds off the darkness. At the same time, he gave us space to cuddle together and weep. He also became the spiritual guide for questions that I was too sad to answer. He recalled verses of inspiration that strengthened our hope and faith as believers in Christ: "We are confident, I say, and willing rather to be absent from the body, and to be present with the Lord" (2 Cor. 5:8). Both the loved ones we lost had accepted Christ, and our hope in their place in eternity had never become more comforting than at that time. What we had taken for granted now became a reality.

Sadly, these losses were just the beginning as we continued each month to lose a loved one to tragedy, Covid, or other illnesses. There were no indoor funeral ceremonies; instead, we drove in funeral processions to the burial sites. "Weep with those who weep" (Rom. 12:15), I would tell my children, and these moments became a lesson in empathy.

As we continued to lose people, sometimes to violent deaths, I would explain that this world is full of human beings with free will.

Sometimes we place too much responsibility on God to protect humanity from tragedy, when the evil comes at the hands of humans. But there are many evils with no clear villain or explanation, like the illness that took so many lives.

As I sought words to explain to my children how a loving God allows such tragedy, I was challenged to look at my own understanding of the Bible. I think of when Job's wife told him to "curse God and die" as they endured losing all ten of their children and everything they owned. His response to her was, "shall we accept good from God, and not trouble?" (Job 2:10).

God never promised us that there would not be trials. What God did promise us is that through it all, he would walk with us and help us find peace in the good and the bad. "Yea, though I walk through the valley of the shadow of death, I will fear no evil: for thou art with me; thy rod and thy staff they comfort me" (Ps. 23:4). I let my children see me pray and cry out to God. Then they also saw me walk in the peace that he gave to me. They were able to experience with me what it means to walk through the valley of the shadow of death and know God's presence. Through this dreadful season, I began to see my children's faith in God grow beyond bounds.

ONE SIGN OF THIS GROWTH came when I lost one of my closest friends. Mr. Coletrane had taught with me in Christian schools for over ten years, including two years as the art teacher at The Living Water School. My children and I have the art they created with him hanging around our home. Suddenly, while driving home one day, he died of a heart attack. My kids and I again found ourselves cuddled in each other's arms, weeping. Then my kids found encouragement in his faith, knowing he was with God. However, I could not get myself together. I just missed him so much. I missed his weekly phone call, where he would greet me, "What's up, Prather?" So full of joy, always with a word of scripture on his tongue. I could not shake the memory of his voice, saying, "Keep pressing, Prather! You're doing the Lord's work!"

I sat there unable to move through my grief. My youngest son, the one who rarely speaks and is the most shy of all my children, came to me with the gentlest voice and asked, "Mommy, did Mr. Coletrane know Jesus?" I looked up and said, "Yes." He continued, "Sometimes when I lose someone I love and I know they knew God, I feel better because I know they are in heaven and I will see them again." He said that and then just stared at me. Instantly, my tears stopped. That year of grieving together was bearing fruit. As we practiced together what it means to walk in faith even in the darkest of times, my children had learned what a life of faith really means: "Now faith is the substance of things hoped for, the evidence of things not seen" (Heb. 11:1). Teaching my children to rest in their faith even when they cannot see hope is the most powerful lesson I could have given them. And in this moment, my youngest son returned the gift to me.

These two years of enduring so much heartache have forced me to abandon any notion that I can shelter my children from sorrow. News of war and images of other families split apart by violence have reminded me of this yet again. I used to think we had to wait to talk to our kids about dying or other tragedies, but these moments of grieving together have taught us to have honest conversations about mortality – even our own. Although I know my children's faith will be tested in this world of so much pain and dark-ness, my hope, my prayer, is that by seeking comfort in God's arms, they will be able to face whatever trials come their way. ➷

Sonnet Addressed to George Oppen, Arlington National Cemetery

I think of the dead, the disposition of
the grave, the marble here arrayed. I've found
my words to be but parodies of sound
or parodies of silence, and (above
all else, perhaps) mere parodies of love.
The 'heartlessness' of words, you wrote (you, bound,
as I), lies in their opacity. We sound
their depths—the *force of clarity*, a covenant. Over by the Mall, the cherry trees
are finishing their dance, and the monuments
are softened by the scent of fading blooms.
Our wounded earth is flooded with a sea
of petals that flick and flutter as they're spent.
Her broad back bends beneath their soft perfume.

ERIC T. RACHER

MAUREEN SWINGER

Stable Condition

An unlikely house extends a welcome to singles and families during Covid and beyond.

All photographs courtesy of Maureen Swinger

Our house in 1998 before its hasty renovation

MY FAMILY LIVES in a barn. If you're picturing an elegant renovation with skylights, open rafters, and cantilevered lofts, that's the one we lived in before this one: woodsy, cozy, just big enough for a family and some guests, snuggled into the hills of northwest Connecticut where Jason and I and our kids spent three years. But upon moving to Fox Hill Bruderhof, eight years ago now, we joined the residents of the large and sprawling former horse barn affectionately named "The Stables."

Our lovely Fox Hill property was originally the West Wind Stables, a horse farm complete with two barns. One is still very much a barn, now sheltering chickens, pigs, goats, sheep, cows, and horses. The other, while looking quite the same from the outside, has undergone dramatic transformations behind its double doors. None of them can be said to have made it beautiful. Not a skylight to be seen.

Let me pause on that insult, and attempt to describe the way housing usually works on a rural Bruderhof. Fox Hill is home to about

Maureen Swinger is a senior editor at Plough. *She lives at the Fox Hill Bruderhof in Walden, New York, with her husband, Jason, and their three children.*

250 people, who live in residential dwellings designed to accommodate four or five families, each in their own apartment. Singles have rooms between the apartments. Kitchens are often shared by two families, as many of our meals are communal and there's less cooking at home.

I've had prospective guests ask me, "What does total community mean, if you share everything? Do you have your own space?" Of course, we respect each other's homes, especially after work hours when the kids need stories and songs, a prayer, and that last "Can I talk about the day?" conversation. We try not to barge in or even phone, so everyone has space to unwind, read, breathe in.

Still, we are a community, and as such, our families extend beyond an immediate circle. There are some amazing singles at Fox Hill who keep the place going, and if any of them need somewhere to chill and talk, or just grill a burger, they know they can stop by our house – if they can find it.

Back to The Stables then, because we live at the extreme end of it, and even frequently visiting friends joke that they need a GPS to figure out the building. They turn up in other people's apartments, confidently – twice. When they try to leave, they walk around our fridge block and appear back where they started with a Hotel California look on their faces.

Since Fox Hill started in 1998, this preexisting building has been adapted to the community's evolving needs, at various times serving as dining hall and communal kitchen, carpentry shop, or elementary school with a library at the heart of it. With each new itera-tion – "Drawers! Floors! Doors!" – not quite as magically as *Encanto*'s Casita, but almost as fast, the interior reconfigured. Eventually, The Stables shape-shifted to family apartments. The old library, once such a cozy den, was demoted to furniture storage. You can't expect folks to live in a big windowless square.

For many of those early years of change, Jason was in the thick of things here at Fox Hill, usually helping with the renovations, and very invested in whichever walls were going up or rapidly coming down. So he wasn't surprised by the look of the place as our family trundled luggage through the dark furniture storage room and down labyrinthine halls to our new abode. But I was a little underwhelmed. Insulated pipes *under* the ceiling panels? Why is this post here, doing nothing, right in front of the kitchen sink? Oh, it's weight-bearing . . . that's all right then. Let's keep the roof up. The floor appears to be polished cement. Is that a footprint polished *into* the cement, here in the girls' room? "Yes," said Jason, "and I think it's mine." He proceeded to step on it, and sure enough, the shoe fit.

It didn't take long for this big, quirky maze to become home. But the furniture room! Every time we walked through, it made its neglect known. The walls were dinged up from clumsy closets, and the light bulbs had burned out in embarrassment. If all the old tables and bureaus went up to join their relatives in the attic, we'd have a central space to gather the thirty-four residents of the building. Suddenly everyone was motivated, patching holes and repainting, putting up pictures, assembling a motley crew of chairs.

We had ourselves a Foyer. And once we had it, we found a flurry of ways to use it . . . not every night, but whenever someone or other said, "How about a get-together?" People contributed board games and books, and soon we were having house game evenings and singalongs, or read-aloud nights with coloring books and crafts in hand.

> **Families and singles contributed board games and books, and soon we were having house game evenings and singalongs.**

As the Christmas season approached, we hauled in a big tree, and a general deck-the-hall afternoon ensued. Then one kid said, "Can we have a fireplace?" Considering the impracticality of punching a chimney up into the living room above, my dad built a faux fireplace, creating the illusion of flickering flames with no heat. All The Stables children hung their stockings along it, and although we don't know how Santa got down, at least the chocolate he bestowed didn't melt.

Gatherings aren't exclusive to Stables residents. The Foyer works for a kindergarten girl's fancy-dress birthday party, a singles' night in, a winter hootenanny, or an annual *Oxford Book of Carols* singalong complete with hot buttered rum and a selection of cherry and mincemeat tarts (the latter an acquired taste, but a critical minority would raise their voices – and not in song – should we forget them).

We learned to grieve together too; when one couple who lived upstairs got the news of their grandson's death in Australia (one of twin boys, his short, brave life had been mostly spent in the NICU), all the residents of the house and several other members of the community gathered at 2:00 a.m. to listen in to the funeral service in solidarity with a family halfway around the world.

WHEN COVID LOCKED the world down, the Bruderhof quarantined in small pods of immediate family and singles, whoever was living in direct proximity and sharing a kitchen. It was the best way in an emergency to protect each other. But going from all-in community to a minimalist, dial-in variety in the space of a day gave all of us whiplash to some degree, especially the kids and the elderly.

Over half The Stables's residents joke about living in "stable condition" because they *are* elderly or health-challenged in some way. And thanks to the Foyer, The Stables could not be divided into small and seemly pods. At once hallway and "great hall," with various kitchens, storage rooms, and exits branching off in hobbit-fashion, it refused to be sealed off tidily, and we were left with two choices. We could become one large first-floor pod of twenty. Or the elderly and immunocompromised could opt to move to smaller, newer buildings, and pod up with close relatives or a caregiver.

It's understood that in a communal life like ours, no one stays in one place forever. You live in the house that works best for you; as a family grows, you might exchange spaces with another who no longer needs as many rooms. If someone has had surgery or is aging and needs more assistance, he or she can move into an apartment with care facilities and attached rooms for nursing staff. But this particular mix in The Stables had stayed together for about five years. Selfishly, I wasn't prepared to see everyone blast apart with no warning.

As it turned out, not a soul wanted to leave. They made up their minds independently (and they wouldn't mind my telling you that they are all recognized for their independent minds), but they let it be known that they considered their mental health on a par with their physical health, and for that, they wanted company. We respected their decisions. In the end, we were fortunate that everyone in Fox Hill kept safe and healthy. But how could anyone have known that at the outset?

Walking home one night with a friend, I looked down toward the long, low-slung silhouette of The Stables, its even line of windows reflecting like small gold portholes in the pond. "Look at it," I exclaimed to my companion, "Doesn't it remind you of the Ark?" just as she helpfully suggested, "the Titanic?"

This was now our community in miniature. While meals were mostly family affairs, literally every other chance to gather was capitalized upon as the pandemic dragged on: old-fashioned entertainments like charades and shadow plays, silly skits, potluck dessert evenings, or themed movie nights. Everyone remembers *The Adventures of Robin Hood*, when the password into Sherwood Foyer was "A Locksley." Come in costume and sneak your ticket past the Sheriff of Nottingham.

It never stopped feeling strange to have worship meetings by pod, each small circle connected yet weirdly remote from fellow members listening in and contributing to the gathering by phone. But we weren't going to give ground to any more distance, any less community.

After the dial tone, folks usually ended up staying on and continuing the meeting more informally, singing (next to impossible while digitally connected), or talking over the day's reading. Other times, we shared stories of pain and sorrow around the circle, perhaps hard news from distant family or friends, or simply an old memory that needed talking through, a grief acknowledged. We made the most of good news, too, and got extra mileage out of birthdays or anniversaries.

THE PODS WERE long ago retired, and since those memorable shared months, some families have drifted out and others flowed in. Right now, an entire extended family has found enough apartments and rooms about the place to gather round their dad and grandfather as he faces advanced cancer, which he does with his signature humor, requests for singalongs, and the occasional ice cream party. The Foyer has become home to civilizations

of block towers built by his small grandsons.

Occasionally, folks from elsewhere come whooshing through when the Foyer is not looking its finest, and an offhand comment will drop: "What a waste of space! Posts right in the middle of it." Or: "No windows. What do you even use it for?" Instantly, I feel an inarticulate defense of all good yet homely things rising within me, as the memories of merry gatherings flash by in kaleidoscopic detail. But probably a fierce and furious justification would just make them wonder about my sanity. I usually end up saying, "Come by at Christmastime." And making a mental note to talk with Bruderhof architects about planning Foyers into future buildings. For sanity. ⤳

Top: a family of musicians serenades The Stables's occupants during the 2020 Covid lockdown. *Bottom:* a communal celebration in 2007.

BRANDON MCGINLEY

Everything Will Not Be OK

You can't protect your children from tragedy.

Nancy Richards
Farese,
Water Play,
Massachusetts,
2018

I TRY NOT TO TELL MY CHILDREN that
everything will be all right.

If they are hurt, I tell them that they
are OK if they are OK, and that a doctor will
fix them up if a doctor needs to fix them up. If
they are scared, I tell them that they are safe if
they are safe, and that I am here for them and
with them no matter what. If they are anxious,
I ask them what they are anxious about, and
we talk through their feelings and reality,
including the reality of their feelings.

But I don't want to get into the habit of
saying, "Everything will be all right." Because
everything will not always be all right.

A friend's boy fell out of a tree. This was a
couple of years ago. His skull cracked on an
exposed root, and he went still. My family
was driving home from visiting relatives out

of state; we told our kids that their friend had had a serious accident, and we prayed and waited through Virginia and Maryland, and into Pennsylvania.

It was important to us that our children, though young (the oldest was about six), knew who and what they were praying for; it increased the emotional burden, yes, but it also enhanced the spiritual communion we all experienced with the stricken family. This was more than a "teachable moment" about tree climbing, though it was that. It was a crash course in reality – the reality that life can change in frightening ways at a moment's notice; the reality that friendship entails experiencing hardship, without hesitation, with our friends; the reality that prayer really is communion, both with God and the people prayed for and with.

Another friend's boy, just this year, was hit by a car. He was being chased by his brother and, in that careless six-year-old way, bolted into the street. The car tossed him far into the air and his head hit the pavement. The family called the parish priest as the ambulance careened to the hospital.

I found out about the accident an hour after it happened, but before anyone knew a prognosis. A mutual friend – the man whose son fell out of the tree – called me as I left the office; as soon as I heard his tone of voice, I knew it was one of those calls, one of those hard days we all dread but anticipate and prepare for. About thirty minutes later, as I

> **It was a crash course in reality – the reality that life can change in frightening ways at a moment's notice.**

prayed on the bus, I checked my phone. A message: the boy had multiple fractures and a concussion, but would be OK.

When I got home, another family was visiting for dinner. We talked about how we knew these days would come, and how relieved we were that this wasn't the big one. We told the assembled children what had happened; my son, the boy's friend and classmate, took it hardest. In that six-year-old way, he couldn't understand and express his feelings, but he was lethargic and short-tempered all evening.

I knelt down and held him by his little shoulders – touch is important in these moments, I've found – and I told him that it was OK, no, *right* to be worried about his friend. I told him that he could pray especially hard for his friend, as the family prayed for him that evening. I told him that the news from the hospital was good – but I did not tell him everything would be all right.

Thus far, we have experienced the most serious traumas in our community only vicariously. But we are intentional about bringing those moments, with discretion, to the entire family. It's part of the communion of community to share the burden of these moments, even with young children.

These moments of uncertainty between normalcy and the resolution of a sudden trauma are the little apocalypses that punctuate every life. They look toward the final apocalypse, the complete unveiling of God's plan for his

---

Brandon McGinley is an author and the deputy editorial page editor for the Pittsburgh Post-Gazette *as well as a contributing editor at* Plough. *He and his wife, Katie, live in Pittsburgh with their five children.*

Nancy
Richards
Farese,
Playing Cars,
Cité Soleil,
Haiti, 2013

creation. They're unpredictable. They're uncontrollable. We have to be ready.

Society is manically anti-apocalyptic these days: driven to avoid or minimize these moments of apocalypse, when a more powerful force of control and causation than ourselves makes Himself apparent. While Americans continue to valorize economic risk-taking in the form of entrepreneurship, risks that more directly implicate day-to-day autonomy are increasingly stigmatized: injury, illness, pregnancy.

This is most clearly true when it comes to children. Challenging, creative play outdoors is discouraged; schools are stupefyingly regimented; fertility is treated like a threat to control rather than a gift to embrace.

It is an indisputable improvement that most people no longer have to suffer the loss

of a sibling in childhood. Child mortality, just under 50 percent of all live births by age five in 1800, in 2020 hit one seventh of one percent in the United States. But that blessed improvement has led to an obsessive impulse to shield children completely from the reality of mortality, from suffering, from all the persistent unpleasantness of being human.

The high child mortality of earlier years was not some aberrant condition. It was not unnatural. There is no violation of some primordial innocence in experiencing the apocalypse of loss or the suffering of a loved one at a young age. It hurts, in every era of human history. It leaves scars. To be sure, it is an evil for children to experience pain in their souls, just as it is for them to experience pain in their bodies. But there are greater evils: the kind of avoidance and sanitization that instills

a false security and a false sense of control, both of which are eventually, inevitably, traumatically destroyed.

When Margaret, our third child and second daughter, was three and a half years old, she drove her face into the sharp corner of a windowsill. The wound next to her left eye was deep but smooth, and only required surgical glue at the emergency room. When we went to our pediatrician for a follow-up visit, he said that if a scar formed and remained in twelve months, we could pursue plastic surgery.

Maybe, if the scar persists, someday Margaret will insist it be repaired. But for now, who cares? A scar is a story, a mark left by life. Of course, some scars of both the body and mind involve lasting pain or serious disability, and are rightly addressed, but there's no need to hide all the evidence of our mortal nature, the evidence of living as a human being – embodied, imperfect, frail.

"My grace is sufficient for you, for my power is made perfect in weakness" (2 Cor. 12:9). It's not in pretending that everything will be all right that we encounter the Lord and his grace; it's precisely in admitting that the trauma of apocalypse is beyond our ability to predict, to control, and to manage that we open ourselves up to him.

The grace of apocalypse is most apparent, and I believe most effectual, when we strip away the veil of euphemism, and regard the life and death offered us and those around us with clear eyes. If an apocalypse is an "unveiling," as the Greek connotes, then at the very least we should hesitate before replacing that veil with one of our own fashioning.

Nancy Richards Farese, *Playing Cars*, Montana, 2012

Nancy Richards Farese, *My Face,* La Gonâve, Haiti, 2017

The goal we aim for in our family is to bear everyday apocalypses not with flinty stoicism but with the faithful confidence that every revelation of God's will, in our time and at the end of time, is also a revelation of God's grace. This is because beyond the greatest trauma in history was the greatest triumph; beyond the cross is the resurrection.

The resilience of the Christian, especially the Christian child, therefore, is not the resilience of the violent who ache to be goaded into conflict, or the suspicious who avoid relationships to insulate themselves from betrayal, or the doomsday preppers who try desperately to game out every scenario. The resilience of the Christian is the knowledge that, in the Lord, everything is always all right – perfectly, serenely, eternally all right – even if here, in the City of Man, it rarely is, and is getting worse.

The global order built by frail human beings is showing its frailty. The self-reinforcing sinews of liberalism and international markets are, it turns out, no match for sin and madness. Conflict hasn't been, and cannot be, completely sublimated to the abstract order of ideology and economics. War is back. It is likely that today's children will inherit a world more violent and more precarious in every way than the one experienced by post–Cold war generations. The belief that everything will be all right was always a recipe for fragility; now it is simply a fantasy.

One of the blessings of living in friendly proximity to dozens of families with dozens of children is that it makes the frictionless, Panglossian view untenable. For all the joy and beauty of the fun days – the impromptu playdates and cookouts, the public and private communal prayer – it's the tearful and frightening days, the apocalyptic days, that teach the most valuable lessons.

We tell our children the truth on those days. We don't worry too much about overwhelming their emotions because the context of these conversations is almost always prayer. "Come to me, all who labor and are heavy laden, and I will give you rest." Children don't have to be the impossibly fragile abstractions of our cultural imagination. They can understand hard things from a young age; they can gaze upon Christ crucified and understand, even if only very dimly, that he shares their burdens; and they can be strengthened by him, even if they can't yet describe what that means.

While children mature at different paces in different ways, usually they can handle reality, including the reality of death, more resolutely than we suppose. Generally it is in treating children as emotionally – and spiritually – brittle that they become so.

During former Pennsylvania senator Rick Santorum's presidential campaigns, an old story often bubbled to the surface: when his little boy Gabriel was stillborn at twenty weeks, Mr. and Mrs. Santorum spent the night with him in the hospital bed, then took him home so their other children could see their brother. Commentators would then imply or say outright that this was unsettling, macabre, creepy.

I wasn't sure that the critics weren't right until I had children of my own. Now I know that, like the Santorums, in that situation we would do everything possible to introduce our children to their brother or sister. The family is the place where we experience the reality of human life, beginning most intimately with cooperating with the Lord in bringing a new person into being. Being open to life means – necessarily, unavoidably, irreducibly – being open to death. It means accepting apocalypse.

When our friends suffered a relatively late miscarriage, they worked with the hospital, parish, and undertaker to hold a proper funeral. At the gravesite, all the neighborhood children tossed earth on the tiny casket. The child was, in a sense, their cousin: Why shouldn't they say hello and goodbye, and participate in the rite dedicating the baby's soul to the Lord?

Even in death, there is grace. That's why we don't feel compelled to say that everything will be all right: Our world doesn't shatter if it isn't. We believe it's held together by sterner stuff than our plans and expectations, and even our hopes and dreams. The Psalmist sings, "A sacrifice to God is a broken spirit; a contrite and humbled heart, O God, Thou wilt not despise" (Ps. 51:17).

The apocalypses of everyday life, little and big, might break bones and spirits. But that's precisely the point. And beyond them the Lord beckons us. He is here, with us; he is there, in the place where everything is unveiled, and everything is healed. ⤚

Potential Space
A Serious Look at Child's Play
Nancy Richards Farese
MW Editions (2021)

This photobook documents children's play across fourteen countries, including Haiti, Cuba, Burkina Faso, Jordan, and the United States.
Farese invites us to consider how this universal activity is threatened by the unrelenting forces of technology, consumerism, and overparenting.

The Spiritual Roots of
Climate Crisis

An Interview with Cardinal Peter Turkson

Plough's Peter Mommsen spoke with Cardinal Peter Turkson, at the time head of the Dicastery for Promoting Integral Human Development, in November 2021 in Rome. In April 2022, Cardinal Turkson was named chancellor of the Pontifical Academy of Sciences and the Pontifical Academy of Social Sciences.

Peter Mommsen: You've played a leading role at the Vatican in calling for action to respond to climate change, pointing out the harms it brings to the world's most vulnerable people. Yet many respond to this reality with indifference or a feeling of resignation. What would you say to them?

Cardinal Peter Turkson: I would say that a more commendable attitude is to learn about and understand what climate change is, and how some of its consequences are already distressingly apparent. There's no question that it's essentially anthropogenic, caused by human conduct and attitude toward creation: Extreme temperatures negatively affect biodiversity and habitat. Unpredictable rainfall disrupts long-standing rhythms of sowing and harvesting, creating and exacerbating hunger. Drought and dwindling icecaps create water insecurity and crisis. Melting ice and rising sea levels threaten island states and coastal settlements. But instead of despair and resignation or indifference, one should be filled with remorse and compunction at how abusively and irresponsibly one treats creation. Isaiah tells us how the earth can languish from sin, and Saint Paul, in chapter 8 of Romans, links the fate of creation with that of its inhabitants. Thus, the proper attitude to adopt in the face of climate change, as Pope Francis observes, is "to become painfully aware, to dare to turn what is happening to the world into our own personal suffering and thus to discover what each of us can do about it."

Somali refugees at a refugee center in Kenya. Climate change was a factor in causing low rain levels in East Africa, contributing to Somalia's 2011 famine.

I can of course understand how some may despair at the enormity and complexity of the phenomenon, and the apparent impossibility of a solution. But indifference in the face of climate change is simply suicidal!

Climate change may be a modern problem, but it's a symptom of an ancient human condition.

It is. While some of its causes are relatively modern – for example, the industrial revolution and the subsequent dependence on greenhouse-gas producing energy sources – the underlying human condition is ancient. It goes back to the experience which scripture describes as the Fall.

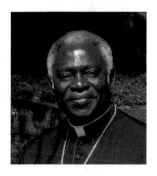

Cardinal
Peter Turkson

In the biblical account of creation, the human person is a relational being, meant to live in ordered and harmonious relationships with God, the creator; with the earth as a garden-home to till and to keep; and with its kind. Climate change could be understood as a symptom of the disordered and exploitative human conduct the scriptures attribute to sin in human life, after the original disobedience to God. The orderly relationship which corresponds to the biblical mandate "to till and to keep" is replaced by activities with little or no regard for the interrelatedness of parts needed to maintain a wholesome balance within creation.

I wonder if you could reflect on the Catholic social teaching of the universal destination of goods; you have called at times for a more communal way of life. Is there anything like this, or other parts of the Christian tradition, that might be brought to bear on such problems as deep poverty and climate change?

Recently, Pope Francis wrote an encyclical letter, *Fratelli tutti*, in which he revisits Saint Francis of Assisi's teaching about how a fraternal bond binds together everything that exists. The friars of Saint Francis were "a band of brothers"; to them, the sun was a brother and the moon a sister. The crucial characteristic of brothers and sisters is that, coming from the same source, they are equal in dignity. Thus the reference to the human family as "brothers all" in the letter's title is a powerful way of affirming the unity and equal dignity of all human beings. Accordingly, the goods of the earth that are meant to safeguard people's dignity and well-being, the "common good," must be destined for all. Vatican Council II puts it this way: "God destined the earth and all it contains for all men and all peoples so that all created things would be shared fairly by all mankind under the guidance of justice tempered by charity."

A powerful synonym of "common good" is what recent popes (Paul VI, John Paul II, Benedict XVI, and Francis) have called "integral human development," the goal or responsibility of a community to ensure the conditions that guarantee the personal, familial, and associative good of its members, so that they may live dignified lives and realize their full integral and personal development.

These sentiments are discernible in the Jubilee Year prescriptions in the Old Testament, and are lived, as a fruit of the Holy Spirit, by the early Christian communities, as described in Acts 2 and 4. The Church Fathers promoted them as expressions of the charity of Christ; Saint Justin includes the taking of collections for the poor in his rite of worship; and Clement of Rome taught that whatever we have is given us by God to be used for building up the body of Christ. The rich should care for the poor in the Church, while the poor should thank God for giving them brothers and sisters able to be generous. But Church teaching also envisages "states and civil institutions that that are primarily concerned with individuals and the common good."

For completeness, we need to mention that the teachings of common good and the universal destination of the goods of the earth are not opposed to private property, business, and the use of capital.

Some economists have responded to this crisis by actually calling for de-growth.

In economic terms, "growth" is a positive: it designates progress and enterprise, for example in the cases of Indigenous communities. In these cases, de-growth would be a negative experience. But when the expression was used by economist Nicholas Georgescu-Roegen in the 1970s, and discussed by the Club of Rome in its 1972 report *The Limits to Growth*, it was intended as a general reminder of the finite nature of the world and its resources, and, particularly, to ask whether some areas could reduce their consumption of resources to enable weaker and poorer nations to advance.

A reminder of the finite supply of everything created is positive and healthy, though it calls on tough ethical muscles to make effective changes. And it suggests solutions to the current search for sustainable development models, for ecological justice and equitable development. Growth that leaves others behind would be against the thrust of sustainable development and calls for some measure of de-growth.

You played a key role in the creation of two major recent encyclicals by Pope Francis: *Laudato si'* and *Fratelli tutti*. Some people charge them with an internal contradiction: they condemn globalized technocracy, while at the same time calling for paths forward that would seem to require even more global technocracy.

Technology is an expression of the talent and creativity of the human mind; and transforming power is what makes us co-creators with God: God created trees, but human technology transforms them into houses and furniture. The industrial and post-industrial ages have been characterized by technological developments which have improved and transformed human life. But the same technology is also poised to dominate, to manipulate, even to destroy the human person. In the words of *Laudato si'*: "Never has humanity had such power over itself, yet nothing ensures that it will be used wisely." Think about its use for medical, economic, and military ends, or about artificial intelligence.

Technocracy suggests that every problem can be solved by technology. This leads to an absolutism that implies that whatever cannot be "fixed" should be discarded, which has clear implications for our understanding of the human person.

As an expression of human creativity, technology serves the well-being of the person, and need not become technocratic. As an exercise of power, technology becomes technocratic, and subjects the person to its rules and interests. Both *Laudato si'* and *Fratelli tutti* are engaged with this important distinction.

> Growth in the sense of progress is something that we strive for. But growth that tends to leave others behind is the beginning of our problem.

You grew up in Ghana. How is your passion on these issues linked to how you grew up?

I was born and grew up in Nsuta Wassaw, a small village where my father worked as a carpenter at the manganese mine. I witnessed how surface-mining methods denude virgin forests of their timber and disembowel the earth with dynamite for the ore that local trains hauled to the Atlantic harbor, Takoradi, about fifty miles away. As children, we blocked our ears with our fingers when the dynamite exploded, and depending on how the wind blew, we would cover our eyes and noses.

Our stream was dammed to form a lake to wash the ore before shipment. To continue to swim in the stream and to fish in it, we had to move upstream, unwittingly invading the ecology of the headwaters.

Within a few miles were gold-mining towns. They had their share of deforestation and gaping holes and caverns from shaft-mining. There was a section of one town everyone called "Cyanide," a sandy part of the town where children played without any sense of what cyanide meant, or of their constant exposure to poison.

Later, as a priest and bishop, I visited several of these mining settlements and villages. Some have become ghost towns; all they have to show are empty, overgrown shafts hidden by weeds and piles of rubble. When a company agent was asked why the shafts are not filled up with the piles of rocks and stones, he responded, "It is not economically feasible." He thought, however, that the gaping holes, contaminated with mercury and other chemicals, could be used as fishponds!

The *New York Times* recently had an article on the phenomenon of young people saying they don't want to have kids because of climate change, or only one . . .

You did ask about de-growth a short while ago, right? The flip side of de-growth is another current of thought, "human ecology," which studies the relations between human beings and their environment. Some exponents of human ecology claim that population increase adversely affects the earth and its resources, and accordingly they advocate for population control to safeguard the well-being of creation. The young people who do not wish to have babies probably consider an increase in population bad news for the earth and the environment.

It is also possible that those who dread the cataclysmic consequences of climate change, instead of committing to promoting eco-friendly lifestyles and habits, are deciding to spare their offspring disaster. I pray that these are not pretexts for irresponsible and liberal ethics about human sexuality and birth. In fact, almost all Western countries are seeing sinking birthrates which threaten the sustainability of their national populations. Governments cannot replace the labor force to support pension schemes. Japan is languishing under anti-birth policies. China has rescinded its one-child policy. And the Western world needs to think about extending its sustainability concerns to population.

It's worth adding that the expression "human ecology" has a completely different application in Catholic social thought and its discourse about the environment. When Catholic social thought speaks about natural ecology, it refers to environmental conditions which are conducive to growth. When it talks about human ecology, it means that humanity also requires a set of conditions (moral, philosophical, economic, health, labor) to be in place for its thriving and successful growth. To paraphrase Pope John Paul II's *Centesimus annus*, for example: Yes, damage to the natural environment is serious, but destruction of the human environment is more serious. We see people concerned about the balance of nature and worried about the natural habitats of various animal species threatened with extinction. But meanwhile, too little effort is made to safeguard the moral conditions for an authentic human ecology.

Integral human ecology as you've just described it doesn't really have political force behind it anywhere in the world. In the United States, for example, you'll have one set of people who are very concerned about the unborn, but fight tooth and nail to prevent environmental regulation – and vice versa.

Disagreements are bound to exist. But it's important that we have principles to guide even

our disagreements, to ensure that we are consistent in our beliefs and actions. Pope Benedict XVI illustrates the point: "It is contradictory to insist that future generations respect the natural environment when our educational systems and laws do not help them to respect themselves. The book of nature is one and indivisible: it takes in not only the environment but also life, sexuality, marriage, the family, social relations: in a word, integral human development. Our duties toward the environment are linked to our duties towards the human person, considered in himself and in relation to others. It would be wrong to uphold one set of duties while trampling on the other. Herein lies a grave contradiction in our mentality and practice today: one which demeans the person, disrupts the environment and damages society." We must insist on the dignity of all persons as images of God and their vocation to brotherhood and sisterhood.

These divides of course are also deep within Christianity itself. But on the positive side, as you were saying, this focus on humans being made in the image of God brings people together. Our meeting is an example: a

Catholic and an Anabaptist. The truth we share draws us together.

In his opening discourse of Vatican Council II, Pope John XXIII said he wanted the windows of the Church to be thrown open, so that the Church can see the reality of the world outside, and the world see into the Church. One reality of the world is its brokenness, as well as the divisions of its churches. As a result, Vatican Council II concluded proceedings with a resolution to establish dialogues with Christian and non-Christian communities.

In humility, we should recognize what happens when we apply human minds to the Word of God. Our interpretations of the Gospels always require the purging of the Spirit. The Lord gave us his gospel and it's meant to go out to the ends of the earth and be put it into practice. Trusting in the presence of the Holy Spirit and submitting to his power to lead us back together, we should confess the Lord and God of our faith in dialogue and in constant friendship. ⬿

This interview was conducted on November 23, 2021; Cardinal Turkson was afterward invited to expand on his remarks in writing. The interview has been edited for clarity and length.

Air pollution from the ArcelorMittal Temirtau steelworks in Karaganda, Kazakhstan

DAVID BENTLEY HART

TRADITION & DISRUPTION

Apocalypse, not dogma, is Christianity's grounds for hope.

ONCE UPON A TIME, Christianity grew and endured and even flourished over the course of many generations in total and blissful ignorance of any officially defined dogma, any single universally recognized canon of scripture, anything remotely like the systematic or dogmatic theologies of the coming ages of Christendom and after. I would add that, for most of that time, there was no single church hierarchy, and that the apostolic lines of succession preserved in later official chronicles were products partly of what one might call retroactive genealogy and partly of what one has to call pious misrepresentations; but we may leave that argument for another time. The point to make here is that, for the first several generations of Christians, anything so precise as a doctrinal symbol authorized by an episcopal council would have been either a curious superfluity of or ponderous encumbrance upon the faith. There had been divisions among Christians even in the apostolic era; the New Testament bears plenteous witness to this reality – so much so

Opposite: Gothic chapel in the Piedmont region, Italy. Photograph by Roman Robroek.

David Bentley Hart is a philosopher, writer, translator, and cultural commentator. He is Templeton Fellow at the University of Notre Dame Institute for Advanced Study.

that the reader can easily get the impression that division was far more common than unity among the early Christian communities. But the principal reason that so many confessional and theological differences of such enormous consequence, on matters so basic to the faith, came to light within the church of the empire only well into the fourth century is that Christian faith and Christian hope had long been sustained by something quite different from official confessional unanimity. The differences had always been there, and in many respects were more or less as old as the faith itself; but for most of the time they were scarcely noticed, since the guiding concern of most Christians was not some perennial wisdom or immemorial doctrine handed down from the past, but rather the rapid approach of the Kingdom of God, the Age to Come, and the final advent of Christ as Lord of all things. Apocalyptic expectation – an eager certainty of the imminence of the full and final revelation of God's truth in a restored and glorified cosmos – and not dogmatic purity was the very essence of faithfulness to the Gospel.

We should therefore never forget that official doctrine is, above all else, a language of disillusionment. The French philosopher Maurice Blondel argued that there must have been more to the eschatological beliefs of the early Christians than the literal anticipation of an imminent Parousia and judgment, as otherwise the faith could not have survived – and with such seeming insouciance – so enormous a failure of expectations. This is a false supposition; and it begs the question of whether indeed one and the same faith did in fact survive. But, putting that aside, surely there should be for Christian consciousness some element of indelible melancholy not only in the thought of doctrinal history's disputes and divisions, but in the very fact of doctrinal definition as such. Doctrine is, in some sense – as much as it may be the poetic discovery of a shared language for speaking about God, and about God and humanity, and about the mystery of Christ – a language of disenchantment that tries at once both to recuperate the force of a cosmic disruption in the form of institutional formulae and to create a stable center within history from which it might be tolerable to await a Kingdom that has been indefinitely deferred. Perhaps this is not to be lamented; at least, a believer has to presume the workings of providence, to the degree that he or she thinks they can be discerned in the midst of fallen time.

Christianity entered human history not as a new creed or system of religious observances, but as apocalypse.

Even so, it should never be forgotten that Christianity entered human history not as a new creed or sapiential path or system of religious observances, but as apocalypse: the sudden unveiling of a mystery hidden in God before the foundation of the world in a historical event without any possible precedent or any conceivable sequel; an overturning of all the orders and hierarchies of the age, here on earth and in the archon-thronged heavens above; the overthrow of all the angelic and daemonic powers and principalities by a

slave legally crucified at the behest of all the religious and political authorities of his time, but raised up by God as the one sole Lord over all the cosmos; the abolition of the partition of Law between peoples; the proclamation of an imminent arrival of the Kingdom and of a new age of creation; an urgent call to all persons to come out from the shelters of social, cultic, and political association into a condition of perilous and unprotected exposure, dwelling nowhere but in the singularity of this event – for the days are short.

The church was given birth in something like a state of crisis, of mingled joy and terror, in a moment out of time, as one age was passing and another coming into existence. The Kingdom was drawing near; the Kingdom had already partly arrived; indeed, the Kingdom was already within, waiting to be revealed to the cosmos in the glory of the children of God. Living thus in history's aftermath, and just on the threshold of eternity, the church could not at first have any expectation that it would soon be required to enter into history again. But it would have to

Chapel in a small abandoned graveyard in the Le Marche region, Italy. Photograph by Roman Robroek.

Abandoned church in the Veneto region, Italy. Photograph by Roman Robroek.

do so eventually, and this meant that it would also have to become everything it thought it had left behind: an institution, a law, a religion. What had begun as an eschatological irruption of eternity into temporal history would in the end – at the far side of a disenchantment so gradual that the initial hope for the imminent Kingdom simply melted, almost unnoticed, into thin air, leaving not a rack behind – have to become just another history: that of a particular creed and devotion and institutional heritage, oriented toward an eternity once again rendered abstract, unimaginable, and inconceivably remote. Soon enough, the church would assume the religious configurations provided by its age, adjusted to accommodate a new set of spiritual aspirations. Jewish scripture provided a grammar for worship, while the common cultic forms of ancient society were easily adaptable to Christian use. After all, a purely apocalyptic consciousness, subsisting entirely in a moment of absolute interruption, could persist for only so long. Still, it was an imperfect synthesis; the alloy of apocalyptic longing and historical continuity was never entirely stable. The Christian event proved to be far too refractory to be contained within institutions, even institutions of its own devising. At the very heart of its spiritual rationale there always remained an impulse to rebellion.

Hence, down the centuries, Christianity has proved not only irrepressibly fissile (as all large religious traditions, to some degree, are), but ultimately self-destructive. Of all the religious cultures the world has ever known, only the Christian has naturally incubated within itself an impulse toward total and defiant faithlessness, militant unbelief, ultimate nihilism, not merely as occasional individual states of soul, but as large cultural movements. Even in its most redoubtable and enduring historical forms, Christianity is filled with an indomitable and subversive ferment, an inner force of dissolution that refuses to crystallize into something inert or stable, but that instead insists upon dispersing itself into the future ever again, to destroy what confines it and to start anew, to begin again in the formless realm of spirit rather than of flesh, of spirit rather than of the letter. There is, simply said, a distinct element of the ungovernable and seditious within the Gospel's power to persuade, one that we ignore only at the cost of fundamentally misunderstanding its most essential character. And this element, with its power to generate intrinsic stresses within even the humblest of Christian communities, could not help but produce a far greater and more chronic stress within the church as an enfranchised institution,

> # Christian dogma has always had some quality of disappointment about it, some impulse to anger.

supporting and supported by the instruments and establishments of a human political authority – an authority now paradoxically allied to a Gospel that consisted to a large degree in the rejection and even damnation of all such instruments and establishments. ("Paradox" is serving here as a euphemism for "contradiction," in case that is not immediately obvious.)

So, as I say, it does not seem foolish to suspect that Christian dogma has always had some quality of disappointment about it, some impulse to anger, some sense that a creed is a strange substitute for the presence of the Kingdom. Dogmatic theology has always had something of the character of a pitched battle among the devout. Perhaps, though, the volatility of theological culture has always been, at some level at least, a reflex of fear: the dread that the truth of the Gospel, exposed to the corrosive force of ordinary time, will dissolve into the currents of an inconclusive history – history without a final cause, and so history without redemption.

The only escape from the desperation this prospect induces is the refuge of tradition understood not as the melancholy memory of a promise that was not fulfilled, but rather as the constant creative recollection of a promise whose fulfillment and ultimate meaning are yet to be unveiled. Tradition thus must be seen as history's secret, redemptive rationale. But tradition of this kind is possible only so long as faith is able to descry a future apocalyptic horizon where the tradition's ultimate meaning is to be found, and is able also to refuse any reduction of that final revelation to whatever formulations of belief happen to be available at any given stage of doctrinal development. If Christian tradition is truly the living thing it must be – at least, if it really is anything more than a collection of accidental associations generated by random historical forces – it must be devoted to that hidden end and not rest content with such dim prefigurations of that end as are already present (and which, as ever, can be glimpsed only in a glass, darkly).

IF CHRISTIAN TRADITION is a living thing, it is only as tradition – as a "handing over," a passage through time, a transmission, the impartation of a gift that remains sealed, a giving always deferred toward a future not yet known – that the secret inner presence in tradition can be made manifest at all. And that gift must remain sealed until the very end, so that the glory will not dissipate into ordinary time, whose atmosphere is incapable of sustaining and nourishing it. The gift is known for now only in and as the dynamic history of the tradition that protects it and bears it onward. Only in the ceaseless flow of the tradition's intertwining variations can the theme subtending the whole music be heard. And in part this is because whatever is imparted must be received in the mode of the recipient, with all his or her limitations and possibilities. In the end, after all, the historical and cultural contingencies of a tradition also constitute the vehicle of its passage through the ages. They are its flesh and blood in any given epoch, its necessary embodiment within the intelligible structures of concrete existence. Without those contingencies, the animating impulse of the tradition would be something less than a ghost. But, by the same token, once that vital force has moved on to assume new living configurations, the attempt unnaturally to preserve earlier forms can achieve nothing but, at the very best, the perfumed repose of a cadaver bedizened by mortuary cosmetics. True fidelity to whatever is most original and most final in a tradition requires a positive desire for moments of dissolution just as much as for passages of recapitulation and refrain. Only in that ceaseless flow of construction, dissolution, and reconstruction is what is truly imperishable in the tradition intuitable.

Alas, there is no single formula for doing any of this well, or any simple method for avoiding misunderstanding. Such rules of interpretation as there are can never be more than general and rather fluid guidelines. They cannot even provide us, when we consult the witness of history, with a dependable scale of proportionality for our judgments upon the past. It is quite possible (and on occasion it has happened) that even the most devout interpreter or community of interpreters, in looking back to the initial moments of the tradition and their immediate sequels and consequences, might reasonably conclude that the overwhelming preponderance of Christian history – its practices, presuppositions, civic orders, governing values, reigning pieties – has amounted to little more than a sustained apostasy from the apostolic exemplars of the church. That hidden source of the tradition's life remains a real and unyielding standard,

Tradition reveals its secrets only through moments of disruption.

and before its judgment even the most venerable of institutional inheritances may have to fall away. And yet, by the very same token, that source remains hidden even within that very act of judgment, and thus can be the exclusive property of no individual or age. Anyone who arrogates to himself the power to say with absolute finality what the one true tradition is will invariably prove something of a fool, and usually something of a thug, and on no account must ever be credited or even countenanced. The claim is in itself indubitable evidence of a more or less total ignorance of the tradition, either as a historical phenomenon or as a dogmatic deposit. And, really, if one is to find the safe middle passage between the Scylla and Charybdis of a destructively pure originalism and a degenerate traditionalism, no particular method can be trusted absolutely; one must instead simply attempt to exercise a certain kind of piety. This requires a certain trusting surrender to a future that cannot alter what has been but that always might nevertheless alter one's understanding of the past both radically and irrevocably. It is the conviction that one has truly heard a call from the realm of the transcendent, but a call that must be heard again before its meaning can be grasped or its summons obeyed; and the labor of interpretation is the diligent practice of waiting attentively in the interval, for fear otherwise of forgetting the tone and content of that first vocation.

In this sense, the living tradition, if indeed it is living, is essentially apocalyptic: an originating disruption of the historical past remembered in light of God's final disruption of the historical (and cosmic) future. One might even conclude that the tradition reveals its secrets only through moments of disruption precisely because it is itself, in its very essence, a disruption: it began entirely as a *novum*, an unanticipated awakening to something hitherto unknown that then requires the entirety of history to interpret. Its abiding truth never suffers itself to be reduced to mere propositional certitudes, but rather testifies to itself in large part by its power to disorder even the temporal forms it has assumed in the course of its pilgrimage through time.

This is the only true faithfulness to the memory of an absolute beginning, a sudden unveiling without precise precedent: an empty tomb, say, or the voice of God heard in rolling thunder, or the descent of the Spirit like a storm of wind or tongues of fire. In a very real sense, the tradition exists only as a sustained apocalypse, a moment of pure awakening preserved as at once an ever-dissolving recollection and an ever-renewed surprise. Any truly faithful return to the origin of the tradition is the renewal of a moment of revolution, and the very act of return is itself a kind of revolutionary venture that, ever and again, is willing to break with the conventional forms of the present in order to serve that deeper truth. What makes the tradition live is that holy thing within that can be neither seen nor touched, which dwells within a sanctuary into which the faithful cannot peer, but which demands their devotion nevertheless. To return to the source is to approach the veil of the Holy of Holies, to draw near once again to the presence on the other side, even sometimes to enter in – though then only to find that the presence remains invisible, hidden in a blaze of glory or an impenetrable cloud. In this way, tradition sets the faithful free.

This essay is adapted from David Bentley Hart's book Tradition and Apocalypse: An Essay on the Future of Christian Belief *(Baker, 2022). Used by permission.*

Jesus
and the
Future
of the
Earth

*To the first Christians,
the age to come was
anything but otherworldly.*

EBERHARD ARNOLD

JESUS BROUGHT FRESH NEWS to the world. It is news about a totally different social order, news that heralds judgment and complete reversal for the life of the present world-age. His news concerns the reign of God, which will bring to an end the present age, the age of man. Without God our age sinks down into hollowness and coldness of heart, into stubbornness and self-delusion. In Jesus, the Father revealed his love to us, a love that wants to conquer and rule everything that once belonged to it. Jesus calls, urging a divided humankind to sit together at one table, God's table, where there is room for all. He invites all people to a meal of fellowship and fetches his guests from the roadsides and skid rows. The future age comes as God's banquet, God's wedding-feast, God's reign of unity. It is a question of God becoming Lord again over his creation, consummating the victory of his spirit of unity and love.

In the Lord's Prayer, Jesus calls on God, our Father, that his primal will should alone prevail on earth, that his age of the future in which he alone rules should draw near. His being, his name, shall at last be hallowed and honored because he alone is worthy. Then God will liberate us from all the evil of the present world, from its wickedness and death, from Satan, the evil one now ruling. God grants forgiveness of sin by making manifest his power and his love. This saves and protects human beings from the hour of temptation, the hour of crisis for the whole world. This is how God conquers the earth, with the burden of its historical development and the human necessity of daily nourishment.

However, the dark powers of godlessness pervade the world as it is today so strongly that they can be conquered only in the last stronghold of the enemy's might, in death itself. So Jesus calls people to his heroic way of an utterly ignominious death. The catastrophe of the final battle must be provoked, for Satan with all his demonic powers can be driven out in no other way. Jesus' death on the cross is the decisive act. This death makes Jesus the sole leader on the new way that reflects the coming time of God. It makes him the sole captain in the great battle which shall consummate God's victory.

There is a gulf between these two deadly hostile camps, between the present and the future, between the age we live in and the coming epoch. Therefore the heroism of Jesus is untimely, hostile in every way to the spirit of the age. For his way subjects every aspect of today's life to the coming goal of the future. God's time is in the future, yet it has been made known now. Its essence and nature and power became a person in Jesus, became history in him, clearly stated in his words and victoriously fought out in his life and deeds. In this Messiah alone God's future is present.

The new future puts an end to all powers, legal systems, and property laws now in force. The coming kingdom reveals itself even now wherever God's all-powerful love unites people in a life of surrendered brotherhood. Jesus proclaimed and brought nothing but God, nothing but his coming rule and order. He founded neither churches nor sects. His life belonged to greater things. Pointing toward the ultimate goal, he gave the direction. He brought us God's compass, which determines the way by taking its bearings from the pole of the future.

Jesus called people to a practical way of loving brotherhood. This is the only way in keeping with our expectation of that which is coming. It alone leads us to others, it alone breaks down the barriers erected by the covetous will to possess, because it is

---

Eberhard Arnold (1883–1935) was the founding editor of Plough *and co-founder of the Bruderhof. This reading is adapted from his 1926 book* The Early Christians in Their Own Words *(Plough).*

determined to give itself to all. The Sermon on the Mount depicts the liberating power of God's love wherever it rules supreme. When Jesus sent out his disciples and ambassadors, he gave them their work assignment, without which no one can live as he did:[1] in word and deed we are to proclaim the imminence of the kingdom. He gives authority to overcome diseases and demonic powers. To oppose the order of the present world epoch and focus on the task at hand we must abandon all possessions and take to the road. The hallmark of his mission is readiness to become a target for people's hatred in the fierce battle of spirits, and finally, to be killed in action. . . .

GOD'S NEW ORDER can break in with all its splendor only after cataclysmic judgment. Death must come before the resurrection of the flesh. The promise of a future millennium is linked to the prophecy of judgment, which will attack the root of the prevailing order. All this springs from the original message passed on by the very first church. There is tension between future and present, God and demons; between selfish, possessive will and the loving, giving will of God; between the present order of the state, which through economic pressures assumes absolute power, and God's coming rule of love and justice. These two antagonistic forces are sharply provoking each other. The present world-age is doomed; in fact, the promised Messiah has already overpowered its champion and leader! This is an accomplished fact. The early church handed down this suprahistorical revolution to the next generation. Jesus rose from the dead; too late did the prince of death realize his power was broken.[2]

From the time of the early church and the apostle Paul, the cross remains the one and only proclamation:[3] Christians shall know only one way, that of being nailed to the cross with Christ. Only dying his death with him leads to resurrection and to the kingdom.[4] No wonder that Celsus, an enemy of the church, was amazed at the centrality of the cross and the resurrection among the Christians.[5] The pagan satirist Lucian was surprised that one who was hung on the cross in Palestine could have introduced this death as a new mystery: dying with him on the cross was the essence of his bequest.[6] The early Christians used to stretch out their hands as a symbol of triumph, imitating the arms extended on the cross.

In their certainty of victory, Christians who were gathered for the Lord's Supper heard the alarmed question of Satan and death, "Who is he that robs us of our power?" They answered with the exultant shout of victory, "Here is Christ, the crucified!"[7] The proclamation of Christ's death at this meal signified the giving of substance to his resurrection, the "doing" of this decisive fact by the transformation and re-formation of life, the giving of reality to that fact of facts which is his victory – born of power and giving power – consummated in suffering and dying, in his rising from death and ascent to the throne, and in his second coming. For what Christ has done he does again and again in his church. His victory is perfected. Terrified, the devil must give up his own. The dragon with seven heads is slain. The evil venom is destroyed.[8]

Thus the church sings the praise of him who became man, who suffered, died and rose again, and overpowered the realm of the underworld when he descended into Hades. He

1. Matthew 10; Mark 6:2–11; and Luke 9:1–6.
2. Ignatius, *Letter to the Ephesians.*

3. Gal. 6:14.
4. Gal. 5:24.
5. Origen, *Against Celsus* VI.34.
6. Lucian, *On the Death of Peregrinus* II.
7. Syriac *Testament of Our Lord Jesus Christ* and the Arabic *Didascalia* (*Didascalia et Constitutiones Apostolorum*, ed. F. X. Funk, Paderborn, 1905, vol. 2), chapter 39.
8. Ode of Solomon 22.

is "the strong," "the mighty," "the immortal."[9] He comes in person to his church, escorted by the hosts of his angel princes. Now the heavens are opened to the believers. They see and hear the choir of singing angels. Christ's coming to the church in the power of the Spirit, here and now, makes his first historical coming and his second, future appearance a certainty. In trembling awe the church experiences her Lord and sovereign as a guest: "Now he has appeared among us!"[10] Some see him sitting in person at the table to share their meal. Celebrating the Lord's Supper is for them a foretaste of the future wedding feast. . . .

The trials of all the Greek heroes cannot match the intensity of this majestic battle between the spirits. By becoming one with the Christ triumphant, early Christian life becomes a soldier's life, sure of victory over the greatest enemy of all time in the bitter struggle with the dark powers of this world-age. No murderous weapons, no amulets, no magic spells or rites are of use in this war. Nor will men look to water, oil, salt, incense, burning lamps, sounding brass, or even to the outward sign of the cross for that mighty victory over the demonic powers, as long as they truly believe in the name of Jesus, the power of his Spirit, his actual life in history, and his suprahistorical victory. Whenever the believers found unity in their meetings, especially when they celebrated baptism and the Lord's Supper and "lovemeal," the power of Christ's presence was indisputable: sick bodies were healed, demons were driven out, sins were forgiven, life and resurrection became certainty, and people were freed from their weaknesses and turned away from their past wrongs.

9. See the oriental and Abyssinian liturgies in F. E. Brightman, *Liturgies Eastern and Western* (pub. 1896), 218. See also the so-called "Clementine Liturgy" in *Apostolic Constitutions*, and the Syriac *Testament of Our Lord*.
10. Quoted by Wetter in Brightman, 452 (the Armenian Liturgy).

HANNA-BARBARA GERL-FALKOVITZ

The Other Side
of Revelation

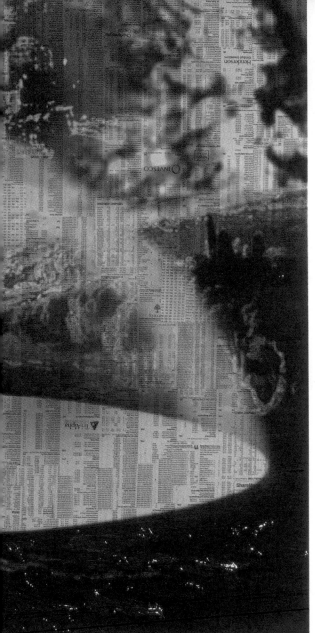

John's Apocalypse can seem terrifying. But that's not how the story ends.

JOHN'S APOCALYPSE IS MOSTLY READ as a book of destruction – and for the first part this is not wrong. But then its vision changes. It presents us with wonderful images of a golden city: of a coming world that is pure light, whose distinguishing feature is overwhelming beauty. The Bible opens with a garden, that is, with human flourishing in nature, but it closes with a city, with human flourishing in culture. Nature is the story's starting point, culture its goal. And that God is seen surrounded by symbols of perfect, captivating beauty – this is the comfort toward which we are journeying.

Hanna-Barbara Gerl-Falkovitz was professor for religious studies at TU Dresden. She now teaches at the Hochschule Benedikt XVI in Heiligenkreuz, Austria.

Gordon Cheung, *Rivers of Bliss*, stock listings, ink, acrylic, gel, and spray on canvas, 2007

The Bride

THE POWERFUL TEXT that appears shortly before the end of the Book of Revelation, in chapters 21 and 22, is like the closing chord of the entire Bible – its finale, toward which everything hurries. More even: everything else

in scripture is bathed in the light of this vision, starting from the beginning of the world. Here the two building blocks are named around which all else is brought together and which together carry the whole edifice. These two elements are the New Jerusalem and Christ, and they come together here in a meeting that has the character of a wedding, of a culminating and final future.

Let us reflect on the first of these two elements: the descent of the city – that is, the bride – accompanied by symbols of penetrating beauty. The image of the bride comes after images of the desolation of the old world, a desolation resulting not from mere abandonment but rather from a willed act of punishment. But once this purification has been completed, the splendor begins. Here we encounter a beauty filled with the power and shimmer of the unimaginable: this city is a perfect cube whose only building blocks are pearls, gold, gemstones, and light. But these are more than just materials: they are emanations of the glory of that light which is the Lamb.

Nothing is only itself; everything is splendor of splendor, a bursting forth of life even from inanimate matter. Here is an outbreak of that same vitality that before had wrought death, shattering all semblances and drowning them in a sea of fire. But now we find nuptial images of life at its zenith, life that awakens no-longer-lifeless matter to an existence full of relationships. For the city (which is simultaneously the bride) is unfurled before our eyes with river and tree, streets and measurable dimensions. Here matter is perfected into light; it becomes transparent. Matter becomes the dwelling place of the bride-church, the place where, at long last, the bridegroom appears.

Poets have taken Revelation's images of perfection and developed them further, seeking to express the inexpressible. In Dante's *Divine Comedy*, the gemstones of the city become a flower, the shining rose of heaven. Here, too, a woman comes into play – Beatrice, who meets Dante at the top of the difficult and steep path through Purgatory and guides him into lightness and brightness. In the words of Romano Guardini: "From the top of the mountain of Purgatory, one floats, or rather, is taken up, in a movement that more and more openly bears the character of being raptured. But what raptures is the smile of [Dante's] guide, Beatrice, she who is wholly beautiful and effortless, the symbol of grace."[1] In Dante's poem, Beatrice represents what we encounter in the Book of Revelation in the symbol of the bride-church. The enchantingly beautiful and effortless way that she guides Dante is summed up by Guardini as follows: "The figure of Beatrice expresses that perfect power does not lie in the greatness of achievement but rather in pure gift, in the smile of the beloved, blessed woman."[2]

Guardini's words allude not only to

1. Romano Guardini, *Vision und Dichtung: Der Charakter von Dantes Göttlicher Komödie* (Wunderlich, 1940), 29.
2. Ibid., 47.

Mary as symbol of the church, but also play on a deep understanding of the divine. For the terms he uses – grace, pure gift, effortlessness – all have to do with divine attributes. "Grace" here may seem a theological term, but like the Latin word *gratia* from which it derives, it evokes something more: beauty. Beauty is among the most glorious and ultimate of God's attributes. As Guardini puts it: "Beauty is the way in which being acquires a face for the heart and learns to speak. In it, being becomes prodigiously loving, and by touching the heart and blood it touches the spirit. That is why beauty is so strong. It sits enthroned and reigns, effortless and staggering."[3] To be sure, the beauty of fallen creation can betray us, twisting the reflection of the divine into evil, so that its expression of perfection has the power to drag us downward – a seduction all the stronger because it still carries the trace of God within itself. One might say that evil has to use beauty to camouflage itself.

Nevertheless, whatever is beautiful still points back to its origin, and Revelation speaks effusively of the eschatological beauty of the redeemed world. Together with this image of the splendor of the bride, John's book calls up one last image to accompany it: that of the bridegroom, the Lamb.

The Lamb That Was Slain

HOW WILL THE BRIDEGROOM be portrayed? The word "lamb" invites us to listen closely – can't you already hear the connotation of slaughter? To keep the splendor from becoming unbearable, even unbelievable, we turn to look at another scene: night in the afternoon, an earthquake, soldiers, and women

3. Romano Guardini, *Religiöse Gestalten in Dostojewskijs Werk* (Kösel, 1947), 256.

and other onlookers at an ancient execution. The man crucified in the center has undergone torture and dies more quickly than the other two. He is marked not only by blood but also by wounds to the head; later descriptions of him will draw on Psalm 22's language of a "worm," of contempt and disgust. It is this executed man who shines in the consummate

city, and he emerges from a horrifying story marked by the pallid light of death and the stench of decay.

By means of this story, the executed man has pulled the world out from the abyss – even if much that is evil and depraved will still come to pass. But he has carried out this rescue at a high price; the Lamb must pay dearly to take on the guilt of the world. For water that washes others will itself become sullied. And this is the drama of Jesus: the utter eradication of the ugly makes God himself ugly. As Paul summarizes this inconceivable process: "He was made sin for our sake" (2 Cor. 5:21). The Lamb himself becomes one of the goats; he allows himself to be steeped in what is hateful and abominable. He becomes not a sinner, but something much more: sin itself. He is unable to seal himself off from his foul-smelling burden; indeed, he becomes indistinguishable from it.

The crucifixion of Jesus is the price paid for our rubbish-ridden and depraved world. Such a world needs more than just a quick external wash. Instead, the Lamb must take on our devouring leprosy. The cleansing he carries out is accomplished with blood and ends in death. Evidently, impurity cannot be overcome in any other way. Impurity is not vanquished from without, but rather taken over from within, in a final show of solidarity. The sacrificial animal hauls itself through the streets of Jerusalem to the place of the rejected, to die there with its burden.

But this act took place once and for all, and, through it, all the filth that we could ever pile up has been taken away. All of it, without exception. There will still be tears, inner aches and inhibitions, and lingering pangs of conscience, but through them the knowledge shimmers: the old has passed away. And far more than that: under the husks of guilt that continue to cling to us, we have become new and different creatures. In the words of Hans Urs von Balthasar, "I have overcome not only death, and not only sin, but sin's disgrace no less, its scarlet infamy, the bitter dregs of your guilt and your remorse and your bad conscience. Look: all this has vanished, leaving less trace than does the snow when it melts under the Easter sun."[4]

∞

Eagerness for the Future

HOW DO WE LIVE for such a future? It's helpful to consider two different ways that the future can be understood. In Latin, time that runs automatically onward from today to tomorrow is called *futurum*. But time that runs in the other direction, from tomorrow to today – time that comes toward me, that shapes me already now – is called *adventus*. This is where we get the word

"adventure," true adventure, the adventure of life.

According to the apostle Paul, the Lord will come "like a thief" (1 Thess. 5:1–6). The announcement of this coming forces existence into a posture of stretching itself out – out toward the coming light, "keeping watch" (1 Thess. 5:6). Such watching and waiting is not dependent on a particular date, but rather is the way a Christian is to live: always already in Advent, always already in the "adventure," doing everything in the present light of the future. What will come tomorrow is already here today. Already today we are redeemed, as will be revealed tomorrow. In this way, the future frees the present from the clutches of the here and now; no longer is the present marked by dull resignation or subjected to fear, no longer is it merely a postponement of the inevitable.

Unlike many other religions, Christian anticipation knows no cycle of rebirth understood as the inescapable, indifferent, and unending primal narrative of downfall and rise, struggle and failure, guilt and dissolution in an eternally repeating, exhausting sequence. Rather, Christian anticipation stands in a history that is urging purposefully forward toward consummation, toward weal or woe. Time is the thought-form characteristic of Christianity: time that sets in motion and keeps in motion. Time is understood as bringing salvation, in a double sense: as *kairos* (the wonderful moment) and *pleroma* (the fullness of time). Within time, what is on offer again and again is new decision: time to be redeemed, time offering itself with an invitation – *my* time.

It is in time and in the flesh that history is realized, pushing forward in the bright flash of decision, not stoically accepted as

---

4. Hans Urs von Balthasar, *The Heart of the World*, trans. Erasmo S. Leiva (Ignatius, 1979), 162.

fate or blind destiny. History is the place in which we are challenged to come face to face with the forces of reality and to endure them. Christianity understands time as the opportunity to be addressed by God and to freely respond to him – with yes, with no, or with evasiveness.

Listening, in the sense of obeying, and refusing to listen, which amounts to becoming deaf, are the two main ways we can respond to the divine claim that confronts us each day. Of course, even when we fail, we may be granted another opportunity, "as long as it is still day" (John 9:4). What's more, any victories we do win must again and again be shored up;

to that extent, the back-and-forth of human willing and doing is a state of suspense that ends only at death. The perfection of the golden city is a victory that can be won only by giving our complete dedication, mind, and will. We must taste the bitterness of the flesh's mortality to the dregs, dying a single and unrepeatable death without any prospect of escape through rebirth, as the wages of sin. Within time, we must fight the battles that will leave us either sanctified or depraved, and must strain under the labor of our yoke, be it the yoke of the Lord or – God forbid – of his adversary. "In hope shall the plowman plow and the thresher thresh" (1 Cor. 9:10).

Gordon Cheung, *On the Horizon*, financial newspaper and acrylic on linen, 2018

Fear – or Joy?

WHAT DOES REVELATION CLAIM about us? What does it say to us?

Apocalypse does not mean simply fear. Rather, it means fear that turns to joy. The old world bursts apart and burns up (today's astrophysics suggests something similar), but then greatness arrives, and no one will mourn

Have we "allowed the flame to die down in our sleepless hearts," in the words of Pierre Teilhard de Chardin? "How many of us are genuinely moved in the depths of our hearts in the wild hope that our earth will be recast? Who is there who sets a course in the midst of our darkness towards the first glimmer of a real dawn?"[5]

It's pointless arguing when and how the goal will be reached: Tomorrow? Or

ερχου κυριε ιησου

the old. It is joy when the old world of sorrow and death is overthrown. For Christianity, this way of escape lies on the horizon of the present transient world. Other religions imagine an endlessly repeating cycle (and endlessly repeating misery). But Christianity waits in confidence for history to culminate in a mighty goal. In earlier times, for the first Christians, this waiting was like a torch in the night.

Israel also waited – for the promised Son of Man, *ben-'adam*. The first time he came helplessly as a child, noticed only by a few. The second time he promises to come openly in power. Power might seem something threatening, but here it announces "summer," just like the young shoots of the fig tree: "From the fig tree learn its lesson: as soon as its branch becomes tender and puts forth its leaves, you know that summer is near. So also, when you see these things taking place, you know that he is near, at the very gates" (Mark 13:28–29). This summer is the summer of a righteous, upright, and rightened existence. For the meaning of judgment is that things will be set right.

far in the future, in the lives of much later generations? One way or the other, the task and comfort of each generation is to remember the ultimate goal, especially in times of fear. In the face of fear, nothing helps so much as the flame of longing for the true Lord of history, for the "heart-bursting delight"[6] of his coming.

The Psalmist looks ahead toward a morning on the far side of death: "As for me, I shall behold your face in righteousness; when I awake I shall be satisfied, beholding your likeness" (Ps. 17:15). In another translation: "When I awake I shall be sated with your beauty."[7] This is the comfort toward which we are journeying. ➤

Translated from German by Cameron Coombe and Peter Mommsen.

---

5. Pierre Teilhard de Chardin, *The Divine Milieu* (Harper & Row, 1960), 152.
6. Thomas Mann, *Joseph und seine Brüder* (S. Fischer, 1964), 1083.
7. Translated directly from the German. In English, a similar approach has been taken by The Psalms Project: *thepsalmsproject.com* —*Trans.*

Editors' Picks

In the Margins
On the Pleasures of Reading and Writing

Elena Ferrante
(Europa Editions)

"Writing with truth is really difficult, perhaps impossible," Elena Ferrante confesses. Her female characters, whom we encounter in moments of crisis, grapple with an age-old tension: the gap between our desire for a story and our imperfect ability to convey in language the truth of what has happened. In Ferrante's globally recognized novels, which include *My Brilliant Friend* and *The Lost Daughter*, this tension is further complicated for women, who struggle to express themselves in a language governed by men.

In the nonfiction essays *In the Margins*, translated from Italian by Ann Goldstein, Ferrante reveals the thought behind her work. At the center of the four essays is "the desire to write," a mercurial force, sometimes "compliant," sometimes "impetuous," with which Ferrante has reckoned since childhood. "The sense I have of writing," she claims, "has to do with the satisfaction of staying beautifully within the margins and, at the same time, with the impression of loss, of waste, because of that success."

Staying within the margins is a metaphor for elegant ("compliant") prose that wins literary prizes. It also reminds us of the historically marginal position of women writers. "Women's writing" emerges here as a contested category, almost a contradiction in terms, where writing has historically been the purview of educated men. For Ferrante, a painful "impression that my woman's brain held me back, limited me" must be continually refuted to produce a literary language capable of expressing female experience. Pain and pen, she argues, are fused together in the "impetuous" attempts of women writers across generations and national borders to move from the margins of literary culture onto the published page.

Alongside writing and "all the struggles it involves," *In the Margins* also models sensitive and perceptive readership. Ferrante's praise for Dante Alighieri in "Dante's Rib," one of many literary inspirations explored in the collection, is a lesson in how the ways we are transformed by what we read need not deter us from gentle and charitable criticism. Without sharing Dante's conviction about Christ as divine, Ferrante acknowledges the arresting beauty of the *Divine Comedy* and, especially, of the poet's muse and guide, Beatrice. As a woman created from man, drawn from Dante's pen as Eve is drawn from Adam's rib, Beatrice becomes an emblem of women's secondary position in the canon. At the same time, she remains a precocious example of "a woman who has an understanding of God and speculative language," a shining tribute to "the gift of speech."

In the Margins addresses longstanding feminist debates around gender and writing, while showing how the problem of representing the world is universal, an "insufficiency of language in the face of love, whether love of another human being or love of God." The final margin to which we must lend attention is therefore a deeply human one: error. This should not depress us, but rather propel us to contemplation, for it tells us that we can always express ourselves better and more fully.

—*Rebecca Walker, PhD,*
University of St Andrews

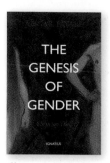

The Genesis of Gender
A Christian Theory

Abigail Favale
(Ignatius Press)

In her new book, *The Genesis of Gender*, Abigail Favale beautifully illustrates a balanced approach to the difficult issue of gender dysphoria. Favale, an English professor and dean at George Fox University, embraced postmodern feminist theory in the early part of her career before being drawn into the Catholic Church by its teaching on Christ's incarnation – and our own. Weaving together personal narrative, academic research, and theological insights, Favale presents a guide through the gender debates that will be particularly helpful to Christians, regardless of their current views on this issue.

After telling her own story, Favale offers a clear and detailed exposition of the creation story in Genesis, contrasting it not only with other ancient creation myths, from the Babylonian *Enūma Eliš* to Plato's *Timaeus*, but also with what she calls the postmodern gender paradigm. In the Christian cosmos, Favale suggests, "sexual differentiation is not a mishap, but cause for celebration and wonder." Our bodies are good, she continues, because "the body reveals the person. . . . Each person's existence is entirely unrepeatable, and our unique personhood can only be made known to others through the frame of our embodiment."

The difference between traditional Christian teaching and today's prevailing gender paradigm centers on their conceptions of reality and its relationship to language. In Genesis, "divine speech makes reality; human speech identifies reality." By contrast, "most gender theories hold that what we think of as 'reality' is a linguistic and social construction." In other words, human speech *creates* rather than *identifies* reality. Building on this key distinction, the middle section of the book presents the intellectual genealogy of various forms of feminist thought, from existentialism to intersectionality, relating and contrasting them not only with each other, but also with the Christian worldview.

The final section of the book is especially powerful. With sensitivity and compassion, Favale shares the stories of men and women who say they have been harmed by gender theory. She emphasizes that "trans identities signal a longing for wholeness, for an integrated sense of self, in which the body does reveal the person. This desire is fundamentally a good one." And yet, while affirming the goodness of this desire for integration, Favale also cautions: "The error comes in thinking that this integration has to be achieved through artifice, through violence against the body, rather than recognizing that we are integrated by our very nature."

Favale calls for Christians to encounter people who identify as transgender as beloved children of God. She tells the stories of people like Addy, whose Catholic roommate patiently listened and asked questions about how Addy reconciled belief in orthodox Christianity with a transgender identity, extending love and acceptance. In the safety of relationships like this, Favale contends, there is room for "not a negation of self, but a rediscovery; not a repudiation of identity, but an unveiling."

As gender norms continue to shift in our society, Favale's book will be an essential guide for Christians who want to encounter their trans-identifying neighbors in a spirit of both truth and love.

—*Serena Sigillito,
Editor-at-Large,* Public Discourse

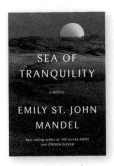

Sea of Tranquility
A Novel

Emily St. John Mandel
(Knopf)

Sea of Tranquility begins on a boat sailing for Canada in 1912 and ends on a space station in 2203. The book is vast, spanning six centuries, two continents, and the moon. Organized around eight seemingly disconnected vignettes featuring a protagonist on a journey of some kind (a second son in pursuit of better prospects, a bereaved wife in search of answers, an author on a book tour, a time traveler), the narrative is a mystery inviting the reader to piece together how these, and indeed all, disparate human lives may share some common themes: connection, death, love.

The book could be described as dystopian. The shadow of plague subtly darkens the micro and macro plots of the book; the scars of smallpox, hints about the 1918 flu pandemic, the specter of the Covid-19 crisis, and a deadly outbreak two centuries in the future. Ideological plagues play their role as well. Colonialism weaves itself into the plot almost unobserved, a quiet continuity from the Canadian appropriation of land from indigenous peoples to the push for colonization of the moon. The portrayal of colonialism is not uncritical, but neither is it activist. Here is one of the many evils that plagues a race for whom nothing is ever enough. When space runs out, time is next.

Yet for a novel that is ostensibly about time travel, pandemics, colonialism, and whether reality is a simulation, the book feels modest, humane, attuned to the particular. The emotional atmosphere of the book is thick with a sense of homesickness, populated by characters separated from those they love by a great distance of either space or time: the older brother, a long-lost friend, the spouse and child waiting at home. In this emphasis on distance there is also a sense of proximity; the theme that pulls these disparate stories together is that desire to be close to those we love, and the battle, sometimes literally, to find each other again when we are lost to time, sickness, or sorrow.

The novel seems to be a work of self-portraiture: Edwin, the protagonist of the first chapter, bears the middle name "St. John," Mirella (of the second chapter) lives through the 2020 pandemic, Olive (of the third) finds herself propelled into fame for writing an (un)timely book about a pandemic, parallel to Mandel's own uncanny composition of *Station Eleven* several years before the Covid-19 pandemic. Looking for herself in every era, Mandel concludes that while pandemics come and go, wars obliterate and remake society, and colonies rise and fall, the pains and joys of human connection persist. At one point Olive observes, "We might reasonably think of the end of the world . . . as a continuous and never-ending process." The time traveler's dilemma (whether to save someone) played out over six centuries reminds us that we cannot escape our own deaths, and yet in moments of hinted synchronicity throughout the book, readers may begin to wonder if we can outlast death after all, through the power that spans these many centuries: love.

—Joy Marie Clarkson,
Plough *Books and Culture editor*

AMERICAN APOCALYPSE

OWEN CYCLOPS

A comic artist explores a peculiar national obsession.

Owen Cyclops is a comic artist and illustrator living in New York City who focuses on history and religion. His first book Channel One: The First Collection of Comics *was released in December 2021.*

ALMOST EVERY GROUP HAD A SLIGHTLY APOCALYPTIC FLAVOR.

NOAH CAME BEFORE THE FLOOD. I HAVE COME BEFORE THE FIRE.

I.E. NEAR THE END

—JOSEPH SMITH, LATTER DAY SAINTS

THIS IS JUST THE ATYPICAL GROUPS, BUT I THINK NORMAL MAINSTREAM AMERICAN CHRISTIANS ALSO HAVE SOME OF THIS FEELING.

THERE'S SOMETHING THAT JUST MAKES THIS COME NATURAL TO AMERICANS.

AT TIMES WE'RE ALMOST PARANOID.

WE'RE A LAND OF PEOPLE WHO ARE OFTEN REALLY RELIGIOUS

IN WAYS THAT OTHERS MIGHT FIND WEIRD.

IT'S A GOOD THING. WE'RE LIKE THE EARLY CHRISTIANS. WE'RE WAITING. WE'RE READY.

JESUS RETURNS SOON

IT'S PART OF WHO WE ARE. IT'S PART OF THIS PLACE. YOU CAN GO OUT INTO A FIELD HERE, AND THERE'S SOMETHING IN THE AIR SAYING, "HEY...

ANY DAY NOW

Syria's Seed Planters

MINDY BELZ

*The war with ISIS spawned a huge wave of refugees.
But not everyone left Syria's Khabur River valley.*

"The years of war have changed the face of the old world. Dynasties and Empires have fallen: old freedoms have been reborn, revolutionary systems of government have arisen. But it is probable that in proportion to its size, no community has undergone trials and upheavals to equal those of the little Nation-Church, which bears the name of the Assyrians." —League of Nations, *The Settlement of the Assyrians, a Work of Humanity and Appeasement*, Geneva, 1935

I TRAVELED TO QAMISHLI for the first time in 2002. Syria then was a sea of state-controlled tranquility in a storm-tossed region. The United States had just gone to war in Afghanistan and would soon be at war in neighboring Iraq. But Qamishli was calm, nestled in Syria's northeast, with a diverse population of Kurds, Arabs, Jews, and Armenian and Assyrian Christians.

The city borders Nusaybin in Turkey, home to one of the oldest surviving churches in Mesopotamia. Qamishli was the gateway to Tur Abdin, or "mountain of the servants of God" in Syriac. Once the apex of the Fertile Crescent, Tur Abdin filled with churches and monasteries in the first centuries of Christian life. From there came early Christian scholarly work, the distribution of Bible texts, and liturgical music.

I flew to Qamishli from Damascus on a quavering Soviet-era passenger plane full of smoking men. As soon as we landed on the tarmac, Kurdish taxi drivers congregated at the base of the boarding stairs to take away passengers. Outside my hotel room window, I heard the market's hum, the mix of Arabic, Syriac, and Kurdish dialects as shoppers clamored over prices. My Kurdish driver picked me up at dawn the next day for an hour's ride to the Iraq border.

We stopped at a roadside restaurant where we took hot tea in paper cups and ate a breakfast of warm flatbread, roasted eggplant, tomatoes, and yogurt. Tributaries of the Tigris and Euphrates rivers irrigate these upper reaches in Hasakah province. The grasslands that day were thick as any carpet, reflecting waves of early morning light caught in gentle breezes.

As a reporter I tend to hurtle toward the next armed conflict, but in this quiet light I sat suffused by the past. Cobbled mounds rose from the plains marking ancient settlements. At least one, Tell Brak, is older than the Egyptian pyramids.

The Khabur River, mentioned in the Old Testament and the writings of the Greeks, carves a valley through this region, which locals call Jazirah. Its ancient people may be the world's first farmers, growing wheat and domesticating wild grasses for livestock. The land's richness transformed nomads to landholders and city people.

Reaching the Tigris before the sun grew hot, I boarded a motorboat to cross into Iraq. Saddam Hussein had turned much of his country into a no-go area for reporters, and

Opposite: Copperplate engraving by Lambert Junior from a drawing by Pierre Jean-Francois Turpin, 1830

Mindy Belz is a writer and former senior editor at World Magazine. *She is the author of* They Say We Are Infidels: On the Run from ISIS with Persecuted Christians in the Middle East *(Tyndale, 2016).*

I entered illegally with help from Kurdish activists and Syria's military intelligence service. With coming wars, first in Iraq and later in Syria, I would not pass this way again until 2019.

Starting in 2011, Syria's civil conflict morphed into an international war. I would manage to enter other parts of the country, but by the time I returned to the northeast, warring parties had fought over it many times: terrorist groups, Syrian government forces, Russians, Americans, Kurdish and Christian militias.

By 2019, Jazirah is a bleakened version of its former self.Islamic State terrorists have captured territory in Syria and Iraq that includes pockets of the Jazirah. The militants control key towns along Syria's northern tier, whose oil and gas fields fund the purchase of weapons. From ISIS headquarters in the northern city of Raqqa, militants run a steady slave trade trafficking women and girls who aren't Muslims, while jailing and abusing any who transgress their radical laws. Slowly coalition forces are driving them out, but scattered cells will remain.

Qamishli somehow escapes ISIS control, but suicide bombings torment its streets. The markets, I find, are nearly empty. Whole blocks go dark at night, a sign of how many residents have left. A government checkpoint divides the city down the center, with Syrian forces controlling one sector while a Kurdish-led force supervises the rest.

The Kurds' alliance with the United States helps wrest areas from ISIS control. The Assad regime in Damascus has more strategic battles elsewhere. Outsiders are beginning to see the Kurd-led governing autonomous administration – with a federal structure and Arab, Kurd, and Christian representatives – as a model for self-rule throughout the country. But war is far from over.

On a rainy night, I skirt the checkpoint area with Assyrian friends to visit Samir Khanoun, a Chaldean Catholic priest. He ushers us into his church's barren reception room, presses to life a diesel heater, and calls for an older gentleman to bring coffee.

Khanoun was born in the village of Tel Baz, south along the Khabur River where the League of Nations created settlements in 1938 for Assyrian Christians who survived genocidal massacres in Turkey and Iraq. About 250,000 Assyrians had been killed. As the survivors flocked to their new sanctuary along the Khabur, Khanoun's parents met, fell in love, and married. Khanoun grew up helping them tend grapes, sheep, cows, and chickens. Each village had at least one church. Worship in the ancient Syriac dialect using centuries-old liturgies shaped Khanoun's daily life. He left Tel Baz for high school in the city of Hasakah, then studied in Egypt and France. "I always was intending to come back," he says.

While he served as a priest in Qamishli, Khanoun's relatives in the Khabur villages came under another attack – this time by Islamic State militants in 2015. The fighters overran thirty-five towns, killing Christian militia members guarding the towns and kidnapping two hundred fifty residents, from six months to ninety years old.

It was one of the global jihadists' largest hostage-takings. Only weeks before, members of the militant group gained international attention for lining up Coptic Christians on a beach in Libya and beheading them. Everyone expected the Khabur captives to suffer the same way.

Seven months later, the jihadists released a video showing three Khabur men dressed in orange jumpsuits, kneeling before hooded gunmen who shot them. Each victim had been born in the villages and was well known

to Khanoun and others. Each had refused a demand to convert to Islam.

The Assyrians had no government to intervene on their behalf, so they turned to the church. For weeks the bishop in Hasakah, Mar Afram Athneil, negotiated with ISIS leaders while funds for ransom arrived from relatives and friends overseas. Over the next year, ISIS released the hostages, first in small groups, and eventually nearly all of them. By then the ordeal had cast a lasting pall over the region.

"The ISIS attack broke all the Christians here," Khanoun says. "They threatened us in Qamishli and everywhere so that people became very afraid." Only a handful of Khabur residents returned to their villages. Daily masses where Khanoun once served more than a hundred people now have "twenty, maybe twenty-five, and the majority are women."

Sudden apocalypse was followed by slow attrition. The 2015 attacks in Khabur and suicide bombs elsewhere in Hasakah province brought violence and cataclysm. Fear brought a different wave of destruction.

Christian families one by one quit Qamishli, Hasakah, and their Khabur settlements. The land that watered and fed their forebears grew fallow. Their flocks ran wild. Their shops were shuttered. Their churches formed smaller and smaller cliques.

Those who had no means to leave or otherwise chose to remain try to carry on amid the slow demise of the life that was. The vendors selling popcorn under evening lights on Qameshli's sidewalks or the clerks scooping falafels behind the counter are small reminders of the way it was before. But Khanoun sees the losses every day in his empty pews and his visits to vacant villages.

"I have offers to leave," he says, "but for me it is a personal decision to stay. To stay until this crisis is over. My Christian faith and responsibility as a priest means I will stay. I want to serve my people even if there's one person left."

The electricity cuts off as we finish. Khanoun insists on serving us again before we leave. He retrieves a large bowl of chocolates in bright wrappers made in a candy factory his

The Church of the Virgin Mary in Tel Nasri on the Khabur River. In 2015, ISIS emptied the town of 1,200 and blew up the church on Easter Sunday. In 2019 only four people had returned.

A young shepherd in the village of Tel Tal

church supports. The chocolates are rich and could be sold in any gourmet grocery. We eat them in the dark.

The next day I head across the flat grasslands again, now west toward the Khabur villages. Though it isn't fully spring, the fields are a blinding green. Our car crosses the old Ottoman railroad that once carried passengers from Baghdad across Jazirah and all the way to Germany. West of Hasakah is a blank expanse before reaching the valley. Cultivated fields, ready to plant, rise into view.

Malik, my translator, explains that in springtime the city families used to picnic along the Khabur riverbank. Now no one will risk it.

We cross the river on a gutted bridge ISIS tried to blow up in 2015. Khanoun told me I'd find nearly all the villages empty. ISIS cells lurk nearby, even after the local forces, helped by US airstrikes, broke their hold on the valley.

Villagers felt abandoned by Kurdish forces in 2015 and still don't trust them four years later. Some Assyrians have moved to larger nearby towns, and some across borders into the mountains of Iraq and Turkey, back to where previous generations began. Others have emigrated to Germany, the United States, or Australia.

Village after village is in ruins. ISIS toppled church steeples, burned and in some cases bulldozed the buildings. It laid mines around homes and watering troughs. It torched grape rows and fruit trees.

As disheartening as it is to walk through the scorched towns, I see the remnants of a pastoral life. Rebar curls from piles of rubble, bullet holes pockmark sheds, but turn the corner and grassy paths lead to neat houses with curtains. Chicken yards sit empty, their fencerows intact. A young boy comes around a lane, leading sheep to a grassy enclosure. Behind the flock his mother shoos them forward, carrying a lamb.

We are in Tel Tal, where I hope to meet Elias Antar, one of the returnees. Assyrians I spoke to from Chicago, Stockholm, and Beirut all implored me to find him if I reached the Khabur.

Antar was born in Tel Tal and plans to die here. He and his wife Shamiram escaped ISIS fighters with many others that February 2015 dawn. The sound of rushing water awakened Antar at around one in the morning. Somewhere upstream, a dam in Turkey that drew down the Khabur for twenty years suddenly had opened. From his living room window Antar watched the flow and couldn't go back to sleep. Soon he heard shots and doors kicked in nearby.

Tel Tal was one of the last villages ISIS reached. Militants arrived as Antar and others escaped to a city upriver. He heard the cries of children and the explosion at the bridge as they went.

Four years later, those days are something he doesn't want to talk about, only to say, "We were the last to escape and the first to return."

When Antar returned to Tel Tal, fighting still coursed through the valley and jets hummed overhead. Too impatient to wait for de-mining teams, he checked for mines himself. His property sits at the edge of Tel Tal and down a winding lane lined with plum trees next to groves of apricots and pomegranates. His fields and fruit trees were torched, his chickens turned out, but he found his house intact. The best way to defeat ISIS, he decided, was to start over.

Now, as years pass and other Assyrians stay away, Antar lobbies old neighbors, relatives, and friends to join his cause. "We challenge all those who are running away by growing things while they are hunting for work in the cities. We have sheep, bees, olive trees, and grapes."

He stands at his front terrace, sweeping his arms wide toward green fields, where women pick mustard greens. He has persuaded a few families to return to the villages. They include the young shepherd and several who had left for America after the attack. He convinced friends from Hasakah and Qamishli to come for day trips to help. Of his seven children, two sons live here. Two daughters emigrated to Germany, two to Australia, and Antar's oldest son moved to Ukraine in 2015.

Four hundred people lived in Tel Tal when ISIS drove them out. Now? I ask. "Maybe fifteen. Twelve or fifteen," he says. "It's difficult to be here," he concedes. "But I work and I have no empty time."

Antar was born in this same house in 1946, he tells me over tea inside his airy, one-story home, its windows open to the Khabur just beyond. His parents' long journey as refugees ended here. His father, forced to flee Turkey, met Antar's mother in Iran. They had to move on to Russia, then Greece, Lebanon, Iraq, and finally to the Khabur settlements, finding sanctuary here. At that time the Assyrians moved in groups, he says, not scattered like now.

In Tel Tal the family farmed cotton and wheat, raised livestock, grew olives, apricots, plums, and more. But "geography is history, and the geography is against us," one of Antar's childhood friends in Chicago tells me. "We are surrounded by many different ethnic groups and most of them are looking at us as *kuffar*," using the derogatory Arabic term for non-Muslims. "It doesn't matter that our people have been there longer."

Elias Antar in his grove of pomegranates

Apocalypse is commonly understood as a complete and final destruction. In its Greek root the term means "to take the cover off," an unveiling. That's why John's apocalyptic book ending the New Testament is called Revelation.

"Some forms of unveiling entail shuttering," author and professor Jeff Bilbro wrote in a 2020 essay, "closing institutions, turning off the lights, going dark. In such darkness, we are forced to stop, take stock, and then learn to go ahead without sight."

I have seen this too many times, where war engulfs civilian populations and ends the world they've known. Many people feel they have no choice but to take flight, forced from their home in the dead of night with no idea where their journey might end.

In 2014 I met Syrian refugees arriving in their pajamas to a snowstorm in Lebanon's Bekaa Valley. Later, I would see Syrians begging on Beirut's streets and living in garages in southern Turkey. I would see them in crowded camps on idyllic Greek isles and in shelters in Paris and Brussels, in transit between a home that is largely gone and a new home they have yet to find.

And always I would meet singular people like Elias Antar and Samir Khanoun, who might serve as fools at the end of a parade but simply couldn't give up. To be at home in Khabur valley, to start over, for all its risks, might be the dream.

Later in 2019, nine months after my visit with Khanoun, I return again to Qamishli. ISIS is fighting what will be its last battle for territory in a town near the Iraq border. US forces want to pull out but instead stay in smaller numbers. Turkey's forces inside Syria increase to create a buffer zone extending along the border. The Turkish forces move downriver, shelling towns and forcing residents from their homes, toward the Khabur valley.

Terrorists are working overtime. Soon after I arrive, a suicide bomb blows up across the street from my hotel. A motorcycle intended also to detonate at the hotel instead runs out of gas and drives with explosives into Khanoun's church. I dial and dial his number with no answer.

I learn he is safe, visiting village churches to the north, and no one else was in his church when the bomb went off. The damage is limited to an outer wall, which he plans to fix.

I stick by my plan to reach Tel Tal and Elias Antar's groves once again, setting out from my hotel as street cleaners clear debris from the bombing. Shops along the street already are reopening, keeping fragile spirits alive.

This time the flat plain is filling with camps for families fleeing the Turkish invasion. The Khabur villages have new residents too, mainly Kurds and Arabs displaced by fighting, who find shelter in the Assyrian Christians' empty homes. Convoys of Russian and American forces move along the main road, and black plumes rise where Turkish forces are shelling old Assyrian towns.

Even so, with my translator and driver I make my way to Tel Tal, and we turn into Antar's lane. I have come to think of him as the mayor of Khabur. He remains unflagging in his campaign for fellow Assyrians to join him, industrious in his labors to make it attractive. Nearby fighting has made no dent in his efforts. In fact, he's taken in a couple from upriver, displaced by Turkish forces, and put them to work tending chickens and rabbits.

As winter nears, Antar is busy with harvest. I find him waiting on his front patio, dressed in a light-colored summer suit, smiling. He beckons me into his groves while Shamiram makes coffee.

We walk through dried grass and wild thyme. He has something to show me just past the plum trees. Four years ago he took cuttings from his charred pomegranate trees and cultivated them into saplings on his patio. He replanted them atop the charred spaces. For the first time, they have fruited.

The pink and red balls are fat and heavy, hanging like baubles on a Christmas tree. He stands proudly by each one, holding them out for show. Then he picks five.

At his patio table we lay them out and he turns over each one. We sip coffee and he fingers prayer beads. "These are the seeds of ISIS," he laughs.

Perhaps what appears lost is only hidden, awaiting its unveiling. Antar seems to possess a secret insight, a hope denied others. What if his Khabur valley's glory days are not only in the past, but waiting in days to come?

Antar pushes the pomegranates across to me: "You will take them to America, and give them to my friends."

And so I do, wrapping each in craft paper on my dining room table and mailing them to the Khabur villagers now living in Chicago.

The Apocalyptic Visions

*As the world careened toward the Great
War, a Russian artist pioneered a path
out of the material to the spiritual.*

SHIRA TELUSHKIN

# of Wassily Kandinsky

T HERE IS A MOMENT, as one rounds the final span of the four-level exhibit of Wassily Kandinsky's works at the Guggenheim in New York, on display through September 2022, where the art gives way to an almost transgressive sense of intimacy.

Previous spread: View of the exhibit *Vasily Kandinsky: Around the Circle,* Solomon R. Guggenheim Museum, New York

The exhibit, *Around the Circle,* is hung in reverse chronological order: We begin at the end, with the artist's final years in 1940s Paris. The show then winds upward through two world wars, the Russian Revolution, his drawings, sketches, the writings on theory and art. As we move back through time, the works start to come into focus, the abstract lines and circles turning back into legible shapes and forms. A house suddenly comes into view as one finally reaches the beginning of the 1910s, his earliest period. A few paces later there are men and women unambiguously conversing over a picnic. Perpendicular to this painting is a train, pumping a recognizable pillar of steam through clearly discernible mountains. We have wound back to Kandinsky's first forays into painting, the start of his experiments with color and texture and light. After nearly seventy works of nuanced spiritual abstraction, these early works hold an almost childlike wonder, so straightforward and requiring no translation. They are still impressionistic, almost dreamlike in their blurred silhouettes and textured brushstrokes of primary colors,

but they feel personal, stripped of those outer layers of meaning and symbolism we have come to expect. It is as if we've intruded on the artist at home, unvarnished, playing around with friends and family, not suited up for serious theological debate.

Having reached the end of the exhibit and the beginning of his life up at the top, I now retrace my steps to the bottom. On this pass, the later paintings seem suddenly sharper, more insightful. The titular painting of the exhibit, *Around the Circle* (1940), stops me in my tracks. These works are viewed, on the return, through the prism of those early works, the directions Kandinsky did not take. I'm not sure the payoff would be the same if the show had started in this earliest era, the paintings growing increasingly bold and abstract to better capture the bold and abstract ideas Kandinsky held about art as a spiritual tool. But by moving backward then forward through time, I can be newly caught by his end, by the strange trajectory his relationship to art and apocalypse and time itself seems to have taken.

Around the Circle is a complex work oriented around a red circle with a purple halo and yellow dot hovering over various other forms set against a dark green canvas. In the bottom right-hand corner there is the outline of an upside-down man, composed from a variety of colors, shapes, and patterns. To his

---

Shira Telushkin lives in Brooklyn, where she writes on religion, art, meaning, and all things beautiful. Her work has appeared in the New York Times, *the* Washington Post, *the* Atlantic, *and many other publications. She is currently working on a book about monastic intrigue in modern America.*

Wassily Kandinsky, *Around the Circle*, May–August 1940

Wassily Kandinsky, *Dominant Curve*, April 1936

left is a door perched above a short flight of stairs, under a collection of bulbous buildings, springing from the crest of a mountain. The sunlike red circle, the man in freefall, the stairs leading up to the door leading up to the city leading back to the sun. The cycle of life. It is an image so quintessentially Kandinsky – the artist's embrace of circles as symbolic of a world which, in his last years, he saw as cyclical, rising and following and moving in never-ending patterns of time and space – that it barely registered when I had first walked in.

Now it captures me.

It sits next to *Dominant Curve* (1936), completed three years after the Russian-born Kandinsky arrived in Paris. With its intersecting circles of orange, yellow, teal, and pale blue, the painting is unmistakably joyful and full of hope.

Who was full of hope in Europe in 1936?

Three years earlier, Kandinsky had fled to Paris from Germany, after the Nazi party shut down the Bauhaus, the influential art school he had helped found in Munich, and threatened to imprison artists, writers, and academics associated with it. And yet in this final stage of his life's work, I see all around me on the gallery floor an energy of gentleness, a move away from his severe geometrical triangles and precise lines to pastel colors and free-floating orbs. Somehow, political turmoil made Kandinsky softer, not harder. Somehow, it was in this period that he became convinced, more than ever, that the world is always evolving and never ending, not marching along a predetermined timeline to some final conclusion.

What an unexpected landing point for an artist who began his spiritual and artistic quest in the 1910s, built on visions of apocalyptic destruction and chaos.

WASSILY KANDINSKY (1866–1944) was born in Moscow but grew up in Odessa, where he attended art school.

He returned to Moscow to pursue law and economics, but a radical encounter with a painting set him on a new course. In 1893, he came across Claude Monet's *Haystacks at Giverny* (1891) in a gallery. He was dismayed: he found himself unable to discern what the painting was meant to represent. He would later write, "This lack of clarity was unpleasant to me: I thought that the painter had no right to paint so unclearly." But the encounter with Monet opened up an important revelation: Art did not have to be simply a reproduction of the world we live in.

For the first time, Kandinsky began to wonder if paintings could not just mimic the physical world, but draw viewers into a higher spiritual consciousness, away from the physical world. Perhaps art could counteract the materialism he saw consuming society, by serving as a portal to divine experiences.

Three years later, in 1896, the thirty-year-old Kandinsky would turn down a prestigious professorship at the University of Dorpat and move to Munich to find his way as an artist.

The path was not easy.

Kandinsky arrived in Munich with nothing but drawings from his school days, and was turned down by nearly every studio and teacher in town. Eventually, with continued practice and persistence, he got himself accepted to a respected studio. Even then, he struggled. He found it maddening to watch artists squander their talent on pretty flowers and nude models, instead of investigating the spiritual potential of their work. In a letter to a friend, he recalled how students would sit for hours in stuffy rooms before an "apathetic, expressionless" model, "trying to represent exactly the anatomy, structure, and character of these people who were of no concern to them," obsessed with the techniques of cross-hatching muscles or shading a nostril or lip, while "they spent, it seemed to me, not one second thinking about art."

Wassily Kandinsky, ca. 1906

Over the years, he would find pockets of inspiration. On a 1908 trip to Bavaria, he became entranced by the local folk art made in glass, depicting saints and Christian heroes with few details but tremendous power. The trip reminded him of an ethnographic study he had conducted in 1889 on the origins of Siberian culture and decoration. The simple, soulful style of the Siberian homes he visited – completely wrapped in bold, decorative art – had struck him as more communicative, more all-consuming, than most of the art one encountered on the walls of European museums. Once again, Kandinsky found that the lack of realism in the folk art of Bavaria allowed the viewer to be transported, as it were, to a realm of the spirit.

In these early years, Kandinsky was most interested in figuring out how to unlock the spiritual possibilities of visual art. A lifelong synesthetic – colors held sounds for him, and he saw colors when he heard words – Kandinsky moved between art forms naturally. He wrote plays and scored librettos. He was exquisitely sensitive to the ways colors interacted and evoked different moods and feelings. He began to seek out writers, thinkers, and other artists who shared his sense of cosmic urgency. Along with friends, he founded a series of societies and magazines. By the start of the 1910s, a burgeoning community of similarly minded people had taken root.

Slowly, he developed a theory of how abstract art could meet its spiritual promise. While to Kandinsky art that merely reproduced the world was worth less than nothing, still for an artist "to limit himself to the purely indefinite would be to exclude the human element and weaken his power of expression."

In other words, images should not be so realistic that they draw a soul down to the material world, but they needed to be recognizable enough to say *something*. For this reason, he was intrigued by imagery that could provide shorthand for complex ideas, allowing works to stay abstract but not forgo all communication. His vividly colorful *Composition VI* (1913), for example, swirls a variety of shades across the canvas with only the suggestion of a hand blowing a trumpet in the left-hand corner, a single clue by which the viewer can access the redemptive message of the painting.

The project felt all the more urgent because Kandinsky believed a crisis was imminent, and artists could either help or do more harm. "Our epoch is a time of tragic collision between matter and spirit and of the downfall of the purely material world," he declared in his 1912 treatise *Concerning the Spiritual in Art*. In this same essay, he develops the idea of history as a spiritual triangle, where the goal of human endeavor is to both move into the upper portions of the ever-narrowing triangle, and to move the triangle as a whole upward. He isn't entirely sure why humans need to keep progressing – he acknowledges that "this need to move ever upward and forward" remains "veiled in obscurity" – but he accepts it without question. This is where art becomes so necessary. Kandinsky believed "painting is a power which can be used to serve the improvement and refinement of the human soul – to, in fact, the raising of the spiritual triangle. If art refrains from doing this work, a chasm remains unbridged, for no other power can take the place of art in this activity."

Given these stakes, Kandinsky bitterly condemned artists who used their gifts to minister to lower needs and "drag back those who are striving to press onward," writing that people looked to art for spiritual elevation, and should not be disappointed. To paint beautiful women or glittering fruit baskets, which might draw people into galleries to say "how nice" or "how splendid" but turn them back out onto the street neither richer nor poorer, was to allow "hungry souls to go hungry away." Art could feed these hungry souls, he believed, and to use it for anything baser was a betrayal of the artistic calling.

This anxiety for the fate of the world, and conviction in the power of art to help or hurt, led Kandinsky to see artists as spiritual leaders, says Richard White, a professor of philosophy who has written about Kandinsky's role in twentieth-century thought. "In Hegel, there are three ways to the absolute: art, religion, and philosophy. Maybe philosophy and religion are not as important as they once were, but art is still so powerful as a way of inspiring people. Some writers talk about how art makes us linger. We look to artists for understanding, and giving us a sense of how the world is. I think Kandinsky saw this as his mission in life."

IN THE EARLY 1910S, now deep into his forties but still in the early stages of his artistic pursuits, Kandinsky began exploring images of apocalypse, destruction, and social collapse, drawing mostly from the biblical Book of Revelation. In this he was not alone. Prompted in part by the Russian loss in the 1904 Russo-Japanese war and the first Russian Revolution in 1905, artists and philosophers alike were animated by "a sense of enormous change and apocalyptic feeling," according to John Bowlt, historian of the twentieth-century Russian avant-garde. Art historian Rose-Carol Washton Long notes similarly that early twentieth-century French and Russian artists, philosophers, and spiritual seekers shared a "messianic vision of a coming utopian epoch."

Between 1909 and 1914, Kandinsky produced dozens of studies with clearly eschatological titles, including *Deluge*, *All Saints*, *Resurrection*, and *The Last Judgement*, in addition to the series of numbered *Compositions*. He experimented with various levels of abstraction and materiality. His 1910 painting, *With Sun*, for example, is done on the underside of glass, and depicts a city of several towers on top of a hill, with waves of fire and water seemingly threatening on all sides. Three horsemen ride up the left side of the painting, while a man and woman lie motionless in the corner just below them.

These scenes of destruction were balanced with an equal interest in Revelation's promise of a new era. In his 1911 *All Saints I*, a figure sounds a trumpet over a rising sun, while a bevy of Orthodox saints emerge with iconography's yellow halos around their heads, led by Saint George in his armor. In the distance is a crucified Christ, and to his left is a new city, encircled by a sunlit wall, resting on a hill.

This line of artistic exploration was cut short by a conflict so destructive it might well have seemed apocalyptic. In 1914, as Europe descended into the Great War, Kandinsky was forced to flee back to Moscow, where he lost ties with his community and friends in Munich. He eventually broke off his relationship with the German-born Gabriele Münter and married a Russian woman, Nina Andreevskaya, who would become central to the direction of his career. Separated from his intellectual circles in Germany, Kandinsky found himself in the new world of the Russian avant-garde. Here he discovered a fascination with mathematics and

"Our epoch is a time of tragic collision between matter and spirit and of the downfall of the purely material world."
—Wassily Kandinsky

the precise geometric shapes that would mark his later work.

In 1922 he returned to Germany to work with the Bauhaus movement, where he painted a series of noisy, explosive works. Forced again to leave in 1933, Kandinsky ended his years in Paris, never abandoning his urgent spiritual desire to create art that draws the human spirit upward, but never again returning to the themes of apocalypse and cosmic redemption that had gripped him before World War I.

THE IDEA OF AN APOCALYPSE – that the present world-age might end, by design, in the foreseeable future – is an idea with a definite historical origin. In the final centuries before the Common Era, the notion that history was a sequence of events playing out in ordered time – a sequence that would one day culminate in some sort of final denouement – began to gain traction across various minority communities in the Levant and Mesopotamia ruled by the Seleucid Empire. We know of no recorded developments of this idea outside the empire.

"It's a reaction formation to the social emergence of a dating system using numbers that never end," explains Paul Kosmin, a professor of ancient history at Harvard who has written extensively about the apocalyptic imagination in this time.

The Seleucid Empire was the first in recorded history to detach its dating system from major events, kingship cycles, and other finite human activities, giving years numbers that could stretch to eternity instead of fixing them to events that would soon be forgotten.

"This new, Seleucid way of counting time generated a crisis of meaning," says Kosmin, "and these apocalypses are the most articulated form of restoring meaning to history. They give it shape, structure."

Up until this point I had only thought of the apocalypse as an ancient idea that felt disturbingly modern because it involved the idea of global destruction. A warming climate, nuclear arms race, and pandemic bring the old themes of war, plague, famine, and death into fresh relief.

Who knew that "crisis of meaning" also deserved to be among this Venn diagram of woes?

"This new sort of linear counting system allowed people to see the kingdom as eternal, or infinite, and it politicized time in a new way," Kosmin continues, adding that this might be one reason we see apocalyptic narratives emerge only among minority groups living unwillingly under the Seleucid Empire, presumably eager to see it end. "The apocalypse makes it possible to oppose the imperial time system, by bringing time itself, which has become eternally forward-moving, to an end."

In other words, the apocalypse as protest.

Kandinsky, who saw the world falling faster and faster into an unending materialistic smog of darkness, would have had much reason to be drawn to theories about the end of time, promising new beginnings.

Eleanor Heartney, an art historian and author of *Doomsday Dreams: The Apocalyptic Imagination in Contemporary Art* (2019), describes a similar impulse today among artists. "The concept of the apocalypse gives history an arc," she said in a 2020 interview with the *Brooklyn Rail*, adding that the apocalypse is not just an imagined end, but a critical way to hope for a new beginning.

When people talk about apocalypse, they usually are thinking of spectacular scenarios of death and destruction, with or without a religious twist. But apocalypse is not just about The End, although that is

Wassily Kandinsky, *Composition VI*, 1913

a big part of it. It is also about justice – the promise that in the end (or actually after the end) good will finally prevail, no matter how awful things seem right now. At the moment, that is a particularly appealing idea. The word *apocalypse* itself means unveiling – it's about revealing the hidden meanings of history. And we want history to have meaning.

But if Kosmin is right about apocalypse's specific historical origin, it is not at all obvious that the desire for history to have meaning is inherent to the human condition. Can it possibly be that the desire for ultimate redemption merely emerged in response to a system of calendars introduced by an ancient empire whose name is obscure to most of us?

> Apocalypse is not just about The End. It is also about justice – the promise that in the end good will finally prevail, no matter how awful things seem right now.

Perhaps. And yet it seems hard to imagine that the apocalyptic texts originating in this period hold our attention simply because they promise some end. They are also, and always have been, dramatically visual. This was the concept from which the biblical books of Daniel and Revelation, as well as non-canonical texts such as the First Book of Enoch, emerged, rich with bloodied angels, lakes of fire, congealed scrolls of ink, and enough beastly creatures to capture the imagination of an artist of any era or creed.

In these texts, Kosmin sees the cryptic, complex symbols as a way to interpret everyday life in the grandest sense. "These second-century Judean texts don't say 'the Seleucid Empire comes out of the sea and is slaughtered,' but 'the fourth beast comes out of the sea and is slaughtered,'" he points out.

Through symbols, the particular becomes the cosmic; the hope for liberation within time becomes the promise of liberation for all time. "Everything that is happening in the apocalypse *to you* is happening *for you*. And you will be the ultimate victor."

It sounds like a project right up Kandinsky's abstract alley. Take the world as we know it and portray it through visuals that are both otherworldly and understandable, clear enough to be interpreted by viewers yet undetermined enough to float free of mundane physical events. Is it any wonder these texts retained such theological and visual power over the millennia, and that Kandinsky found in them the tools he needed to tempt the deadened souls of his own generation towards deeper spiritual reflection?

IN RECENT DECADES, there has been a trend to subsume Kandinsky's interest in apocalyptic motifs wholly within the broader interests in theosophy and spiritualism present in early twentieth-century Europe. Such influences likely did shape the artist. Theosophists of the time believed folklore and mythology were ideal ways to communicate spiritual concepts to the masses; similarly, the very accessibility of the Revelation motifs might have been one reason Kandinsky turned to the book of Revelation. He was known to express his admiration for Helena Blavatsky, the co-founder of the Theosophical Society in 1875, and even more for Rudolf Steiner, a leader of Christian theosophy who was teaching the Book of Revelation in Munich at the same time Kandinsky lived there. The links to his art seem clear.

But such theories too easily overlook the influence of Kandinsky's pious Russian Orthodox upbringing and the long tradition of iconography that goes with it.

Kandinsky grew up during the revival of the Russian Orthodox Church's tradition of iconic images in the late nineteenth century. In Orthodoxy, icons are often seen as guiding objects to help one grow in spiritual awareness, useful to gaze upon during prayer. They are rife with symbolism: the suggestion of a lion, a wheel, or a cloak can instantly conjure up a world of religious themes and emotions for someone educated in the church's symbolisms. For this tradition, the choice of colors matters immensely, and images are not primarily a narrative or depiction of a scene, but rather a kind of divine ladder to the godly realm.

It's a short leap from this way of thinking to Kandinsky's experiments in abstract art, and one that he was better equipped to make than his Protestant-raised colleagues. As we've seen, his attraction to the abstract arose from his desire for art to be a portal to the divine, not a narrative scene, which tends to keep the viewer in this world rather than drawing him or her beyond it. Connecting icons with Kandinsky's turn to the abstract doesn't seem so far-fetched, given how strongly Kandinsky expressed his sense of Christian identity in personal writings, even if he rarely articulated confessional beliefs in public.

One riddle remains: Why did Kandinsky lose interest in apocalypse? How did the man drawn in his pre-1914 years to images of the destruction of the world later end up promoting visions of a spiritual cycle of time with no definitive end or beginning?

As I wind through his works in the Guggenheim show, I find myself thinking a clue may lie in the religious motifs on which he so often drew throughout his life. It can be easy to forget that Kandinsky was many things before he was an artist. He was a son, a husband, a lawyer, a child of the Russian Orthodox Church; a seeker and a thinker, a seeing ear and a listening eye. Writing in 1913 to a friend, Kandinsky mused about the similar avenues of rupture and rebirth that religion and art offer:

Art in many respects resembles religion. Its development consists not of new discoveries that obliterate old truths and stamp them as false (as is apparently the case in science). Its development consists in moments of sudden illumination, resembling a flash of lightning, of explosions that burst in the sky like fireworks, scattering a whole "bouquet" of different colored stars around them. This illumination reveals with blinding clarity new perspectives . . . the continuing growth of earlier wisdom, which is not canceled out by the latter, but remains living and productive as truth and as wisdom. Christ, in his own words, came not to overthrow the old law, [but] he transformed the old material law into his own spiritual law. In this way, I have since come to conceive of nonobjective painting not as a negation of all previous art. I have always been put out by assertions that I intended to overthrow the old tradition of such painting.

The end was never an end, for Kandinsky, but always a transformation. Not an overthrow of the world but a sublimation into something new. Even after he moved on from his preoccupation with the Book of Revelation, he retained his determination to keep moving upward, even if by the end of his life he wasn't sure there would ever be a final resting stop.

For all his frustration with other artists, Kandinsky never wavered in his belief in the power of art to be a spiritual catalyst. In his words: "Despite all this confusion, this chaos, this wild hunt for notoriety, the spiritual triangle slowly but surely with irresistible strength moves onward and upward. The invisible Moses descends from the mountains and sees the dance round the golden calf. But he brings with him fresh stones of wisdom to man." ⤵

JOEL CLARKSON

At the End of the Ages Is a Song

To prepare for Christ to return and bring all things to completion, the early Christians sang.

IN THE DAYS PRIOR to Russia's invasion of Ukraine, as fear built to a froth in news reports, military projections, and NGO preparations, something strange happened: people began to gather, often in public, to sing. In a video captured for the world to see, a crowd of Ukrainian Christians stood in a subway entrance in Kyiv and sang out the spiritual anthem "Prayer for Ukraine" as others passed in a frantic rush before war hit the city. Those of us watching from a distance may have registered cognitive dissonance between the portent of cataclysm and the strangely peaceful crowd, seemingly unaffected by the chaos around them. But for those Ukrainian Christians in the subway, singing was more than mere catharsis or distraction. It was an entry into a posture of expectancy and hope, an engagement with a practice of the early church that many of us have forgotten: the belief that the whole of the created order is made from music, and that to sing is to prepare for the end of the world.

Early Christians expected Christ to return soon and restore the world. The Incarnation initiated a new relationship between world and God, and in Christ's paschal sacrifice on the cross the curtain of the temple was torn asunder. The rites and rituals reserved for the holy of holies now seeped out into the world, and every activity within creation, from the patterns of prayer to the movement of the sun and moon, became imbued with liturgical significance. Robert Taft writes that in the worship of the early church, "all of creation is a cosmic sacrament of our saving God. . . . For the Christian everything, including the morning and evening, the day and the night, the sun and its setting, can be a means of communication with God."

To prepare for Christ to return and bring all things to completion, Christians sang. Some of the earliest hymns reflect this cosmic motif. An example of this is the Oxyrhynchus hymn from the late third century, the earliest instance of Christian hymnody with accompanying musical notation. Though the text has only been maintained in fragments due to the manuscript's tattered quality, the surviving passage is striking: ". . . all splendid creations of God . . . must not keep silent, or shall the light-bearing stars remain behind . . . All waves of thundering streams shall praise our Father and Son, and Holy Ghost, all powers shall join in . . ." For the first followers of Christ, the whole of creation became a call to praise, and every act of praise became a participation in the Parousia, the second coming. Even as the first Christians realized that the return of Christ would not necessarily be in their own lifetimes, they retained a sense that the second coming would not occur at a future time, but was an ongoing event in the here and now. Jesus was coming quickly; even creation

Opposite: Praying woman, Catacomb of Priscilla, Rome AD 200–400

Joel Clarkson is a composer of film, TV, and sacred music. He is writing a doctoral thesis on theology and music through the School of Divinity at the University of St Andrews in Scotland.

The Oxyrhynchus hymn, on a third-century papyrus scrap, is the earliest Christian song for which we have musical notation.

groaned in expectation, as Romans 8:22 says. To sing was to groan with it.

And there was reason to groan. Christians faced constant threat of persecution and death, and were painfully aware that life was fleeting and time was short. Whether Christ returned, or they found themselves suddenly in his presence due to martyrdom, to these first believers the present age, with all its tribulations, was passing, and the time of the kingdom was near at hand. Nature became a living metaphor of this passage of one age to the next, and the progressions of natural light at dawn and dusk took on an eschatological potency. Cyprian, a third-century bishop of Carthage martyred under Valerian, makes this connection in *Treatise on the Lord's Prayer*, instructing Christians on the necessity of praying at dawn and dusk:

> For since Christ is the true Sun and the true Day, as the sun and the day of the world recede, when we pray and petition that the light come upon us again, we pray for the coming of Christ to provide us with the grace of eternal light.

Exemplifying this practice is one of the most celebrated hymns of Christian tradition, and one of its earliest: the "Gloria in excelsis Deo." Drawing on the angelic song to the shepherds in Luke 2:14, this vibrant acclamation, which in time became an integral aspect of the Mass, began as a song praising God for the rising light at dawn. Amid the shadows of affliction and even annihilation, these patterns

of prayer, set in song and saturated with the significance of creational illumination, pointed to the One in whom hope might truly be located.

The fourth century introduced a new, complicating dimension into Christian life. With the Edict of Milan in AD 313, Christianity embarked into the uncharted waters of widespread societal acceptance. After decades of the most intense persecution in the history of the church, suddenly Christianity was not only legal but desirable as a religious identity. Christian faith entered the mainstream, and with it, all the patterns of eschatological hope underwent a seismic shift. No longer did the former urgency, precipitated by persecution and death, exist to press Christians into expectant prayer. In its place, a new peril set in: the lull of passivity. How was the church to instill the anticipation made so real in the daily prayer of the first three centuries?

One answer came in the proliferation of monastic communities, which adopted an ascetic lifestyle, often in remote locales, and formed "rules of life." These communities continued the practice of daily prayer as a discipline of their vows, under the influence of the fathers and mothers of monasticism such as Anthony the Great, whose life came into profile in the Western church through Athanasius's *Life of Saint Anthony*; and Basil the Great, whose monastic settlement in Cappadocia established a precedent of community life, prayer, and mysticism continued to this day, especially in the East. For the vast majority of Christians, however, this unique vocation would remain out of reach. How was the church to respond to the challenges posed by living in an empire replete with worldly vices and material excesses?

Alongside monastic practice, recent scholarship has identified another tradition that scholars such as Taft call a "cathedral office" of prayer, which would have been intended for the laity. Unlike monasticism, which established a pattern of obligatory prayer at set hours throughout the day, this cathedral office centered on dawn and dusk. Like its precursors, it engaged in daily prayer as expectation of Christ, and saw him present in the sign of the rising and setting sun. Vespers, the evening liturgy, expressed this in a unique way. At the hour of dusk, as the shadows set in, a lamp would be brought into the congregation to be blessed. The practice, called the *Lucernarium*, was a long-established custom throughout the Mediterranean world, in various religious and cultural persuasions. And yet Christians transformed the convention into a potent sign of worship: in that flame they recognized Christ, the true light of the world, whose light would, in time, replace even the sun itself. And the response to that sign, the means through which Christians might participate in that light and make its meaning real in their lives, was to sing.

The most common example of this was the hymn of lamp-lighting in the East, the *Phos hilaron*, which remains in use in Orthodox vespers liturgy today:

O joyful light of the holy glory of the immortal Father, the heavenly, holy, blessed Jesus Christ. Now that we have reached the setting of the sun and behold the evening light, we sing to God: Father, Son, and Holy Spirit. It is fitting at all times to praise you with cheerful voices, O Son of God, the Giver of life. Behold, the world sings your glory.

In singing this hymn, Christians made themselves part of the living sign of the *Lucernarium* lamplight, entering the now-and-soon-to-be of its eschatological hope.

Christ lit the way not only in persecution and in comfort, but also in death. In a climactic moment in his biography of his beloved sister, Macrina, Gregory of Nyssa describes her at her

Macrina the Younger, the sister of Basil the Great and Gregory of Nyssa (ca. AD 327–379).

bedside, where, even as she enters the twilight hours of her life, she sees and responds to the *Lucernarium*:

> Meanwhile evening had come and a lamp was brought in. All at once she opened the orb of her eyes and looked towards the light, wanting to repeat the thanksgiving sung at the Lighting of the Lamps. But her voice failed and she fulfilled her intention in the heart and by moving her hands, while her lips stirred in sympathy with her inward desire.

For Macrina, Gregory seems to say, the hymn of lamp-lighting, said at dusk, prepares her to enter the dawn of her resurrection in heaven.

> ## In song, the signs of Christ's coming continue to shine brightly for those who have eyes to see, ears to hear, and lungs to sing.

From the other side of the Mediterranean comes a strikingly similar story from another beloved saint, this time concerning the *Deus creator omnium*, a popular fourth-century Western vesperal hymn comparable to the *Phos hilaron*, written by Ambrose. In his *Confessions*, Augustine recounts his intent not to publicly grieve his mother Monica's death, even as his sadness leaves him feeling shattered. In this unrest, Augustine recalls Ambrose's hymn:

> The bitterness of my grief exuded not from my heart. Then I slept, and on awaking found my grief not a little mitigated; and as I lay alone upon my bed, there came into my mind those true verses of Thy Ambrose, for Thou art –

> Deus creator omnium,
> Polique rector, vestiens.

Recalling these words, Augustine finally weeps, falling into the comforting and compassionate arms of God. In reflecting on the hymn of vespers, Augustine is able to accept grief as the right response to death, seeing God as the one in whom all endings in the brokenness of creation's twilight, including death, are not truly endings.

Within less than two centuries, the stability won in those early decades of the fourth century would be ripped to shreds: the empire, sinking into decay, would be destroyed, Rome would be sacked, and once again, Christians would be forced to contemplate the shadows of the present age and the hope of Christ's return. Looking back, it is clear how important it would be for those Christians to keep that hope alive.

In our time, too, the shadows of war and illness and injustice still lurk in the corners, even as creature comforts, technological advances, and prosperity seek to distract us and make us complacent. How are we to remain alert and ready for the coming of Christ?

If we attune our ears, we can hear it in the chant of the earliest Christians, singing psalms as they were led to martyrdom; we can hear it in the hymn of lamp-lighting sung by the Cappadocians during the fourth century; and we can hear it in our own time in Ukrainians singing their prayerful song in the face of imminent calamity. To sing as a Christian isn't to deny or avoid the fallen realities of the world in some sort of escapism; rather, it is to enter into the midst of them, and to declare that though the darkness may seem strong, a light shines in the darkness which the darkness cannot comprehend (John 1:5), and which, in the fullness of time, will banish the darkness for good. In song, the signs of Christ's coming continue to shine brightly for those who have eyes to see, ears to hear, and lungs to sing.

PLOUGH BOOKLIST

New Releases

With or Without Me: A Memoir of Losing and Finding

Esther Maria Magnis, translated by Alta L. Price

With or Without Me is an unsparing and eloquent critique of religion. Yet Esther Maria Magnis's frustration is merely the beginning of a tortuous journey toward faith – one punctuated by personal losses retold with bluntness and immediacy. Magnis knows believing in God is anything but easy. Because he allows people to suffer. Because he's invisible. And silent. "I think we miss God," she writes. "I would never want to persuade anyone or put myself above atheists. I know there are good reasons not to believe. But sometimes I think most people are just sad that he's not there."

Robert Spaemann: I have not known anyone since Nietzsche who shows so shockingly what a catastrophe it is to not believe in God.

Softcover, 201 pages, ~~$17.99~~ **$12.59 with subscriber discount**

Following the Call: Living the Sermon on the Mount Together

Eberhard Arnold, Augustine, Wendell Berry, Dietrich Bonhoeffer, Dorothy Day, Meister Eckhart, Timothy Keller, Søren Kierkegaard, Martin Luther King Jr., C. S. Lewis, Richard Rohr, Dorothy L. Sayers, Rabindranath Tagore, Barbara Brown Taylor, Mother Teresa, Leo Tolstoy, N. T. Wright, and ninety-four others

Edited by Charles E. Moore

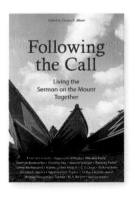

Jesus' most famous teaching, the Sermon on the Mount, possesses an irresistible quality. Who hasn't felt stirred and unsettled after reading these words, which get to the root of the human condition? This anthology is designed to be read together with others, to inspire communities of faith to discuss what it might look like to put these radical teachings into practice today.

Russell Moore, *Christianity Today*: This book will prompt you to surprise, to delight, to melancholy, to argument, and, at every turn, will lead you back to Jesus.

Softcover, 396 pages, ~~$18.00~~ **$12.60 with subscriber discount**

Brisbane: A Novel

Eugene Vodolazkin, translated by Marian Schwartz

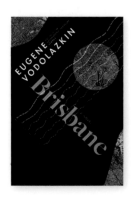

In this richly layered novel by the Ukrainian-born Russian author of *Laurus*, a celebrated guitarist robbed of his talent by Parkinson's disease seeks other paths to immortality.

This personal story of a lifetime quest for meaning will resonate with readers of Dostoyevsky, Tolstoy, Umberto Eco, and Solzhenitsyn. Expanding the literary universe spun in his earlier novels, Vodolazkin explores music and fame, belonging and purpose, time and eternity.

Hardcover, 343 pages, ~~$26.95~~ **$18.87 with subscriber discount**

Vision for a Time of Crisis

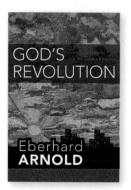

God's Revolution: Justice, Community, and the Coming Kingdom
Eberhard Arnold
Introduction by Stanley Hauerwas

Feeling powerless to change the greed and injustice at every level of society? Tired of answers that ignore the true causes of human suffering? This revised anthology of Arnold's most compelling writings challenges us to seek the eternal truths of Christ's way. But be warned: to Arnold, discipleship means revolution – a transformation that begins within, but spreads outward to encompass every aspect of life. Here is the raw reality of the gospel that has power to change the world.

Thomas Merton: Arnold's writing has all the simple, luminous, direct vision into things that I have come to associate with his name. It has the authentic ring of a truly evangelical Christianity and moves me deeply. It stirs to repentance and renewal. I am deeply grateful for it.

Hardcover, 204 pages, ~~$20.00~~ $14.00 with subscriber discount

Action in Waiting
Christoph Friedrich Blumhardt

Given the number of people who've been "saved," you'd think the world was becoming a brighter place. In his quest to get to the essentials of faith, Blumhardt burns away the religious trappings of modern piety like so much chaff. His active expectation of God's kingdom shows us that the object of our hope is not relegated to some afterlife. Today, in our world, it can come into its own – if only we are ready.

Eugene H. Peterson: The life and writings of Christoph Blumhardt are adrenaline for faltering and compromised followers of Jesus.

Softcover, 253 pages, ~~$12.00~~ $8.40 with subscriber discount

Seeking Peace: Notes and Conversations along the Way
Johann Christoph Arnold

For anyone sick of the spiritual soup filling so many bookstore shelves these days, *Seeking Peace* is sure to satisfy a deep hunger. Arnold offers no easy solutions, but also no unrealistic promises. He spells out what peace demands. There is a peace greater than self-fulfillment, he writes. But you won't find it if you go looking for it. It is waiting for everyone ready to sacrifice the search for individual peace, everyone ready to "die to self." *Seeking Peace* plumbs a wealth of spiritual traditions and draws on the wisdom of some exceptional (and some very ordinary) people who have found peace in surprising places.

Softcover, 254 pages, ~~$12.00~~ $8.40 with subscriber discount

(continued from p. 120)

a convent of the Poor Clares. Meanwhile, he spent long hours contemplating the Gospels and meditating on Christ's presence in the Eucharist. This form of devotion had always deeply moved him. To think that God would make himself so small as to offer himself up as a humble piece of bread!

Eventually, he came to the realization that he could live in the manner of the Man of Nazareth anywhere – and that he must do so where he could be of greatest service to others. Not surprisingly, in accordance with his "logic" of the Incarnation, he was drawn to the poorest of the poor. He longed to be a brother to them – to assure those whom the world despised that God had not forgotten them, and to share their fate.

In 1901, de Foucauld returned to the Sahara and the Tuareg, a tribe that lived on the margins of the world known to Europeans at that time, with virtually no knowledge of Western civilization or Christianity. He would spend the next fifteen years with them, accepting their bare-bones standard of living and offering hospitality and care to the poor and sick. He also immersed himself in their culture, learning their language and studying their traditions.

De Foucauld often dreamed of calling other brothers to join him in his work – he envisioned founding a community modeled on the one that sprang up around Jesus and his disciples. But this dream was not to be realized, at least not during his lifetime. For one thing, his clerical superiors, unsettled by his tendency to "absolutize" ideals such as self-denial and the renunciation of family ties, dissuaded him from recruiting followers. For another, the handful who joined him anyway soon discovered that his ascetic lifestyle was all but inimitable. As the old saying goes, "A saint is a wonderful person, as long as you don't have to live with him."

Charles de Foucauld, ca. 1907, Algeria

On December 1, 1916, against the backdrop of regional unrest generated by World War I, de Foucauld was kidnapped and shot by bandits. And yet his legacy lives on. De Foucauld's memory was kept alive by a Catholic association he had helped to organize, and popularized by a bestselling 1921 biography. Communities formed that sought to promote his theological insights and put them into practice. While the first of these groups arose in his native France, others have emerged elsewhere in the intervening century, among them the Little Sisters and Little Brothers, also known as the Communities of Charles de Foucauld. Inspired by his life, and what he called the "spirit of Nazareth," they choose to live in voluntary poverty among the poor, and to serve the "hidden Christ" as he is found on the margins of society – in prisons, depressed urban areas, refugee centers, and other places of despair. ⇲

Translated by Chris Zimmerman from Wer alles gibt, hat die Hände frei: Mit Charles de Foucauld einfach leben lernen *(Droemer Knaur, 2021).*

Charles de Foucauld

*A young French hedonist follows Jesus
into the desert.*

ANDREAS KNAPP

*Opposite:
Cory
Mendenhall,
Charles de
Foucauld,
watercolor,
2022*

Artwork by Cory Mendenhall. Used by permission.

DECLARED A SAINT BY POPE FRANCIS in May 2022, Charles de Foucauld (1858–1916) was the scion of a long line of wealthy French noblemen. After the early death of his parents, he followed in the footsteps of his forebears, as expected of him, and embarked on a military career. Influenced by a milieu that held a dim view of religious belief, he also spent lavishly, burning through his inheritance and becoming notorious for his dissolute lifestyle. Pleasure-seeking did not satisfy him, however, and he was dogged by an inner emptiness. A tour of military duty and a subsequent geographical research expedition in the Algerian Sahara (at that time a part of France's colonial empire) turned him – to the astonishment of his peers – from party animal to desert fox. Back in Paris, he was welcomed as a dashing adventurer; inwardly, he was churning. What had happened?

For one thing, he had witnessed Muslims praying in Morocco – a sight that deeply impressed him and now drew him to the church. Fortunately, the clear-sighted priest to whom de Foucauld turned recognized that the young man in front of him was not facing an intellectual dilemma but an existential decision. In the course of ongoing conversations, the priest invited de Foucauld to confess his sins and put his life into God's hands.

De Foucauld was deeply moved by this experience and later spoke of it as a vital turning point in his life. In line with his tendency to commit himself to something completely or not at all, he sought entry to a Trappist monastery in Syria – a poor community in a famously strict order. As he explained the decision to Henri Duveyrier, the well-known explorer with whom he had traveled North Africa: "Why did I join the Trappists? Out of love – pure love. I love our Lord and cannot bear to live any other life than his. I cannot travel first-class while the one I love went through it in the last."

A few years later, after witnessing the massacre of local Syrian Christians on the orders of the Turkish sultan (he and his fellow French monks were left unharmed), de Foucauld found himself longing for a different way to follow Christ. If God, in his love for humankind, had taken on the form of a man and chosen the downward path of poverty, shouldn't he – in solidarity with that man, Jesus – stand with the poor and powerless too?

In 1897, leaving the security of the monastery, he moved to Nazareth, in Palestine. Here, in this unassuming village, he sought to come nearer to an understanding of the Incarnation – the mystery of a God who had chosen to live as a simple craftsman among ordinary people as they went about their daily lives.

Pleased to be so close to his beloved Lord, both geographically and inwardly, de Foucauld found work as a house servant at

(continued on p. 119)

Andreas Knapp is a member of the Little Brothers of the Gospel in Leipzig, Germany.